T0191909

More-than-Moore 2.5D and 3D SiP Integration

Riko Radojcic

More-than-Moore 2.5D and 3D SiP Integration

 Springer

Riko Radojcic
San Diego, CA
USA

ISBN 978-3-319-84932-4   ISBN 978-3-319-52548-8   (eBook)
DOI 10.1007/978-3-319-52548-8

Printed on acid-free paper

This Springer imprint is published by Springer Nature
The registered company is Springer International Publishing AG
The registered company address is: Gewerbestrasse 11, 6330 Cham, Switzerland

Preface

There is a lot of buzz in the industry about More-than-Moore technology directions, and especially the 2.5D and 3D System in Package (SiP) integration options. In principle, More-than-Moore integration, and 2.5D and 3D SiP implementation, is an opportunity that can be leveraged to extend system level miniaturization without More-Moore type of scaling, and to bring incremental cost-power-performance value. At a superficial level, it would appear that the SiP concept of integrating multiple die in a package—either by putting them side by side (2.5D integration) or on top of each other (3D integration)—is pretty straightforward, and not all that novel. This would seem especially so now that the key enabling foundational technology modules, such as the Through Si Vias and the uBumps, have been proven out, and are in volume manufacturing. After all, there are several announced SiP products that leverage the 2.5D and 3D integration technologies, and more are rumored to be coming. So, what is the big deal?

However, in practice, implementation of competitive 2.5D and 3D SiPs for mainstream products—especially for cost conscious consumer market—is challenging. The entire IC product design and sourcing ecosystem, and the standard industry practices and methodologies, have all been optimized over the last few decades for sourcing 2D SoC type of products, and are therefore challenged by some of the requirements for, and uniqueness of, 2.5D and 3D SiPs. Thus, adoption of 2.5D and 3D technologies is disruptive to the standard industry paradigms.

The intent of this book is to explore the tradeoffs that need to be considered in order to make 2.5D and/or 3D integration technologies attractive for use in high volume IC components targeting consumer products, competing in, for example, the mobile market. The new degrees of freedom offered, as well as the new constraints imposed by these integration technologies are in fact quite insidious. The tradeoffs required to optimize a 2.5D or 3D SiP products and make them competitive versus traditional 2D SoC IC are complex, mutually interdependent, involve architecture, design, Si and package technologies, have consequences in multiple physical domains, and impact both the technical and business considerations. It IS complicated!

This book therefore reviews the various popular technology options for both 2.5D and 3D integration, with focus on the candidates that are productizable within a 3- to 5-year implementation horizon. Basic manufacturing process flows, and the associated supply chain, required to realize a competitive SiP product are summarized. 2.5D SiP integration options based on Si, Organic Substrate, or Glass Interposers, or on Fan Out Wafer Level Package technology, are all reviewed. 3D SiP integration based on Through Si Via chip stacking, including the various technology and stacking options are presented. The value propositions of 2.5D and/or 3D integration are outlined, with a focus on monetization opportunities in high volume consumer marketspace. The potential system-level benefits of tighter memory integration—either via an HBM stack in a 2.5D package or WideIO in a 3D package—are presented. The opportunity of improving component-level cost structure through splitting a mono-die SoC into a Split Die SiP are explored. The 2.5D and 3D SiP architecture, physical design, Si and Package process technology, and product sourcing tradeoffs and considerations are explored, and the effects of the differences between 2.5D/3D integration versus 2D SoC are amplified. The upgrades to the design methodologies and EDA tools required to optimize 2.5D and 3D SiPs are identified, and some practical short cut solutions are proposed. The impact of 2.5D and 3D integration on electrical performance as well as on the thermal and mechanical stress characteristics, are described, and methodologies for addressing the potential interactions are outlined. Typical standard practices used in making new IC product development decisions, as well as the structure of the typical corporate entities involved in sourcing IC products, are also reviewed, and the business implications of adopting a disruptive More-than-Moore type of technology are assessed.

Thus, this book provides a holistic perspective that spans the broad process and the design technologies and methodologies, and addresses both, the technical and business considerations. It is not quite an 'everything-you-always-wanted-to-know-about-2.5D/3D-technology-but-were-afraid-to-ask' type of work, but it does elaborate on why adoption of More-than-Moore type of 2.5D and 3D integration technologies is complicated. It can be done, but it is complicated…

San Diego, USA   Riko Radojcic

Acknowledgements

The learning described in this book is a result of an effort to evaluate the 2.5D and 3D integration opportunities for semiconductor products competing in the consumer mobile market. The effort was a multi-company collaboration that spanned several years and involved several excellent forward looking multidisciplinary engineering teams. As such, it is not possible to name all the companies, teams or individuals who have contributed in different and significant ways—but a big "thank you" is due to all. However, the author would like to recognize the outstanding effort, and talent, of the Advanced Technology Integration team at Qualcomm, who were at the center of these 2.5D and 3D evaluations. Furthermore, this work would not be complete without naming a few colleagues, nor would it be fair to ignore the outstanding contributors and teachers amongst them. So, a special kudos and thank you is owed to Matt Nowak, Sam Gu, Durodami Lisk, Brian Henderson, Urmi Ray, Mark Nakamoto, Anup Keval, Maxime Leclercq, Rajiv Dunne, Karthikeyan Dhandapani, Wei Zhao, Juzer Fatehi, Magesh Govindarajan, Ilya Gindentuller, Dong Wook Kim, Jae Sik Lee, Ron Lindley, Miguel Miranda, Chandra Nimmagadda, Dan Perry, Vidhya Ramachandran, Thomas Toms, Martin We, Sherry Wu... Thank You, one and all. It has been an honor and pleasure.

Contents

Definition of Acronyms

ASP Average Selling Price
AUC Average Unit Cost
BEOL Back End of Line
BGA Ball Grid Array
BLR Board Level Reliability
CFD Computational Fluid Dynamics
CPI Chip-Package Interaction
CTE Coefficient of Thermal Expansion
CVD Chemical Vapor Deposition
D2D Die-to-Die
DDR Double Data Rate
DFT Design for Test
DRM Design Rule Manual
EDA Electronic Design Automation
FCBGA Flip Chip Ball Grid Array
FEM Finite Element Model
FEOL Front End of Line
FOM Figure of Merit
HBM High Bandwidth Memory
HMC Hybrid Memory Cube
HVM High Volume Manufacturing
I/O Input-Output
KGD Known Good Die
KoZ Keep out Zone
L-o-L Logic-on-Logic
LPDDR Low Power DDR
LVM Low Volume Manufacturing
MCM Multi Chip Module
MEOL Middle End of Line
MM More-Moore
M-o-L Memory-on-Logic

MtM	More-than-Moore
OSAT	Outsourced Semiconductor Assembly and Test
P/G	Power and Ground
PCB	Printed Circuit Board
PDK	Process Design Kit
PDN	Power Distribution Network
PI	Power Integrity
PoP	Package-on-Package
RDL	Re-Distribution Layer
SAP	Semi-Additive Process
SI	Signal Integrity
SiP	System-in-Package
SoC	System-on-Chip
T2T	Tier-to-Tier
TCB	Thermal Compression Bonding
TDDB	Temporary Bond and De-Bond
TGV	Through Glass Via
TMV	Through Mold Via
TSV	Through Si Via
UBM	Under Bump Metallization

Definition of Terms

- "**Technology**" is a generic term, but here often refers to manufacturing process technologies, including Si wafer processing, assembly and packaging processing, and the associated design enablement.
- "**Incumbent Technology**" refers to technologies that are an evolutionary derivatives of an existing mainstream technology used in current products in an existing market, and with an easy projected value proposition. For example, scaling CMOS technology from node n to node n + 1, along the More-Moore path, is an incumbent technology solution. Similarly pushing package substrate technology from 10 to 8 um line/space spec purely through improvements in process and materials is an incumbent technology solution. And so on.
- "**Disruptive Technology**" refers to technologies that break a trend in technology evolution and/or roadmaps and require a change in product architecture and/or the design to generate a value proposition. For example, TSV-based 3D stacking is a disruptive technology solution. WideIO memory is disruptive relative to LPDDR memory. And so on.
- "**Value Proposition**" refers to the balance of the cost-benefit equation, and the analyses of the various elements that contribute to the cost and/or benefit for a product is referred to as tradeoff assessment. i.e., conceptually: Value = (Benefit) − (Cost).
- Terms like "Strata" or "Levels" or "Floors" have been used in the industry to refer to the individual levels in a 3D stack. Here the term used is "**Tier**". The bottom die is referred to as Tier 1 (T1), next is Tier 2 (T2), etc.
- Term "**Die-to-Die**" (D2D) is used to denote die interconnections for 2.5D side by side integration and "**Tier-to-Tier**" (T2T) is used to denote die interconnections for 3D stacked integration.
- "**Partitioning**" refers to different ways of distributing existing functions in a system (e.g., CPU, GPU, PHY, Memory, etc...) across a die or a system. Conventionally a system contains for example a Digital die (with CPU, GPU, Modem, etc...), DRAM, a non-volatile Memory, some display drivers, a PMIC, etc. Partitioning studies looked at different way of distributing these functions across different dice.

Chapter 1
Introduction

Semiconductor Industry—a ~ US$300B + business—is at crossroads. For the last 50 years, i.e., since the beginning of its time, it has prospered and proliferated in most amazing ways, but pretty much following a single paradigm, which is described by Moore's Law, as illustrated in Fig. 1.1.

Moore's Law—the very well-known empirical observation that the number of transistors per chip doubles every 18–24 months, first articulated by Gordon Moore in 1965, is nowadays exemplified by the modern system-on-chip (SoC) chips with literally billions of transistors per die, enabling the miracles of internet, smart-phones, PC's, mainframes, and all sorts of smart appliances that only a few years ago were the realm of sci-fi.

This has been achieved mostly through "scaling"—where the minimum dimensions of features on a chip realized during manufacturing are reduced in each successive generation of process technology—resulting in the doubling of the number of transistors in approximately the same die size, at an approximately constant cost. Of course, along the way, many new device types were introduced, the number of the levels of interconnect has ballooned, many new materials were invented, and process techniques that appear to defy the laws of physics were developed.

This general paradigm is coming to an end—for both, technical and economic reasons. With feature sizes down to a few nanometers and approaching atomic dimensions, it is increasingly difficult to achieve dimensional scaling, and with the number of masking steps approaching 100 masks, the cost of processing a Si wafer is ballooning.

However, there is no expectation by the industry that the evolution of technologies and the increasing levels of integration at the system level, and the miniaturization at the appliance level, will cease. This is an ongoing trend that has historically transgressed multiple classes of technologies, as described, for example, by Kurzweil (Kurzweil 1990; Kurzweil 1999).

General expectation in the industry is that the next technology paradigm that will extend system miniaturization and integration will be through the so-called

© Springer International Publishing AG 2017
R. Radojcic, *More-than-Moore 2.5D and 3D SiP Integration*,
DOI 10.1007/978-3-319-52548-8_1

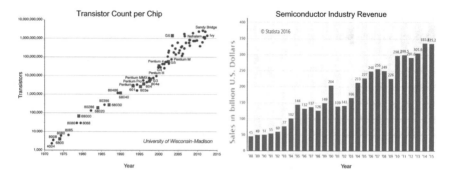

Fig. 1.1 Moore's Law Virtuous Circle—a plot showing Transistor count per leading edge CPU chip over last ~45 years (Fig. 1.1a), and Semiconductor Industry revenue for the last 25 years (Fig. 1.1b)

"More-than-Moore" (MtM) (Arden et al. 2016; Zhang and Roosmalen 2009) class of technologies—versus the "More-Moore" (MM) technology class exemplified by straight pursuit of Moore's Law. Whereas the portfolio of "More-than-Moore" technology options is not really clear at this time, broadly these technology options could be segregated into three types; namely:

i. *Embedded Si Solutions*: those options that are implemented purely through in-fab/foundry processes—for example—by embedding new device types on a monolithic Si die, such as MRAM or RRAM embedded memory

ii. *Embedded Package Solutions*: those options that are implemented purely through packaging processes—for example—by integrating various passive components in a single package, such as Integrated Passive Devices (IPD), or Passives-on-Glass (POG) implementation, and

iii. *System-in-Package Solutions*: those options that leverage a combination of in-fab and packaging processes—for example—by integrating multiple die on top of each other using through Si vias (3D TSV stacking), or by assembling multiple die side by side in a single package (2.5D integration).

This book is focused mostly on the third class, i.e., the "System-in-Package" (SiP) implementation—in contrast to the "System on a Chip" (SoC) implementation that is typically associated with the More-Moore path. The reason is that the first class of technology options, that embed new device types in Si die, is quite similar to the traditional More-Moore class of technologies, and shares similar constraints and design tradeoff options. Similarly, the second class of technology options, that integrate passive elements, is an alternative and potentially better way of implementing R-C-L networks, and requires no tradeoffs at the Si chip level. The third class, not only exercises both the Si foundry and OSAT packaging type of process technologies, but also requires tradeoffs in design domain in order to achieve an optimal implementation, and as such it is believed to best represent the More-than-Moore opportunities and challenges.

Current expectation is that the SiP type of products will be realized through the new integration technologies, often referred to as "2.5D" and/or "3D" integration. These technologies leverage suitable combination of in Si and packaging processes and focus on integrating higher density of functions per unit volume, and/or lower cost per function, at the package, PCB, and/or system level—as opposed to increasing the density of transistors at a die level.

2.5D and/or 3D integration technologies are sometimes perceived as an easier path than continuing the More-Moore paradigm down to 7, 5, 3 nm… CMOS nodes—a path that requires revolutionary innovation in the process and/or material and/or device architecture domains. On the other hand, 2.5D/3D integration techniques are also perceived as disruptive to many of the existing practices, requiring a new design ecosystem. Hence, it is not clear whether, or when, SiP approach may displace SoC's, or if and how the two paradigms will coexist.

The intent of this book is to present the current state of the art of the More-than-Moore/System-in-Package integration technologies, including their potential applications and value propositions, and the associated technical tradeoffs. The book addresses the design and architecture tradeoffs, and the potential implications for the entire product sourcing ecosystem, as well as reviewing the status of the Si and packaging process technologies. In addition, typical business practices used to assess the overall readiness of 2.5D/3D technologies are reviewed. That is, the book is intended to offer a holistic review of the challenges and requirements for "prime time" use of 2.5D/3D technologies in commercial, high volume, consumer products.

This book is segregated into three basic sections, as follows:

SiP Technology Opportunities—Chaps. 2 and 3 describe the various candidate 2.5D (Chap. 2) and 3D (Chap. 3) stacking integration technologies intended for SiP solutions, and summarize the current state of the art of the various process options, along with the associated value propositions and technical tradeoff parameters. Whereas, clearly not every single possible SiP implementation is exhaustively described, the section does identify the most common candidates, and highlight the current challenges and tradeoffs required to meet the competitive product performance, cost, yield, and reliability specifications.

SiP Design Challenges—Chap. 4 describes the methodology and tool challenges and opportunities, and the associated tradeoffs, for design of 2.5D/3D SiP products. For successful mainstream (as opposed to niche application) commercial deployment, an entire product design ecosystem, including the architecture, design, test, characterization, and the associated tools, may need to be updated—in addition to developing the manufacturing process technology.

SiP Product Landscape—Chap. 5 describes the typical practices used in the industry to weigh a new technology paradigm, and to assess both the technical and business sides of a product implementation decision. The focus is on the dominant sector of the semiconductor industry that competes in the advanced digital SoC market with leading edge CMOS technologies. This sector pretty much shaped the technology directions and the standard industry practices during the last 50-odd years and accounts for the majority share of the overall semiconductor industry revenue. Since adoption of SiP product options for volume applications will be weighed versus an incumbent SoC solution, understanding of the landscape of this industry sector is necessary to understand the SiP versus SoC decision. Also included in this section are some subjective projections and opinions, based on author's experiences.

The book is written for readers who participate in the semiconductor industry and are familiar with the usual terminology and basic principles of IC integration, Si and packaging technologies, etc.… It is targeted for an individual, or a team, who is considering the benefits and costs of More-than-Moore integration.

References

Kurzweil Ray (1990) The age of intelligent machines. MIT Press, Cambridge, MA. ISBN 0-262-11121-7
Kurzweil Ray (1999) The age of spiritual machines: when computers exceed human intelligence. Penguin Books, New York, NY. ISBN 0-670-88217-8
Arden W et al (2016) More-than-Moore White Paper, http://www.itrs2.net/uploads/4/9/7/7/49775221/irc-itrs-mtm-v2_3.pdf. Accessed 24 Nov 2016
Zhang GQ, Roosmalen A (eds) (2009) More than Moore. Springer, US, ISBN 978-1-4899-8431-9

Chapter 2
More-than-Moore Technology Opportunities: 2.5D SiP

2.1 Overview

More-than-Moore (MtM) technologies and associated practices are currently not clearly defined—it is more of a concept for continued miniaturization of electronic systems than a tight definition of established practices and requirements. The technology direction is probably still best described by the concept chart published in 2005 edition of ITRS (International Technology Roadmap for Semiconductors 2005), and illustrated in Fig. 2.1.

There is plenty of discussion ongoing in the industry—for example under the umbrella of the ITRS 2.0 and its successors—to define the scope and requirements for this class of technologies (Gargini 2016; International Technology Roadmap for Semiconductors 2.0 2015). However, it is generally agreed that at least one of the essential elements of the More-than-Moore technology vector is to increase system-level integration by leveraging packaging technologies to integrate multiple similar (homogenous integration) or dissimilar (heterogeneous integration) die in a single package. This set of technologies are then referred to as 'System-in-Package' or SiP solutions.

The fundamental preposition of this work is to present the technology options, their value prepositions, and the design tradeoffs required to make SiP solutions a viable alternative to the traditional SoC implementation for high volume commercial appliances.

There are two generic types of SiP technologies:

 i. multiple die in a single package placed side by side. This is sometimes referred to as "2.5D Integration"
 ii. multiple die in a single package stacked on top of each other using Through-Si-Vias. This is referred to as "3D Integration"

© Springer International Publishing AG 2017
R. Radojcic, *More-than-Moore 2.5D and 3D SiP Integration*,
DOI 10.1007/978-3-319-52548-8_2

**Fig. 2.1 More-Moore
versus More-than-Moore
paths**—a concept chart that
defines the More-Moore and
More-than-Moore integration
strategies (from ITRS 2005)

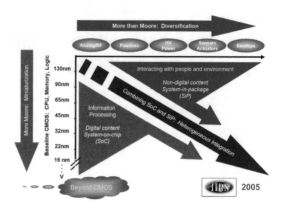

This chapter is focused on 2.5D Integration Technologies. In general, placing multiple die in a single package reduces the length of the interconnect between the components, the system component count, and potentially, the overall system form factor. Reduced interconnect length, and reduced component count typically results in reduced system power and enhanced performance. Eliminating a package by housing multiple die in a shared package can also reduce the overall cost. Thus, the concept of multiple die in a package is, and always has been, attractive to system engineers. In addition, housing multiple die in a single package allows semiconductor vendors to bundle several die types into a single product, and/or to match the characteristics of the die to enhance performance or yield, thereby potentially increasing their market share. Thus, the concept is attractive to component engineers, as well. Hence, the idea of integrating multiple die in a single package is not at all new. In the 80s and 90s these were referred to as Multi Chip Modules (MCM). Even earlier (circa 1960s) "Hybrid Modules" often included multiple die as well as several passive components, usually on a single ceramic substrate.

The difference of the modern incarnation—is that this implementation uses higher density of interconnect between the die—typically of the order of 1000s of wires—than is typically practical with multiple single chip packages, thereby enabling use of a more parallel interfaces. Parallel interfaces can be used to slow down the clocks and reduce power while increasing data throughput, and potentially to simplify the design of the I/Os. Thus, modern 2.5D Integration is more than just several chips in a package, in the sense that it enables differentiation at the architecture and design level—something that is typically hard to equal with standard single chip package implementation. This requirement for the higher density interconnect, however, pushes the packaging technologies beyond the current state of the art driven by the standard single chip packaging requirements—with implications on manufacturing cost, yield, and throughput.

The sketch in Fig. 2.2, illustrates the generic, conceptual 2.5D implementation and highlights the key differentiating features:

  i. *Interposer*: a separate layer in the package that contains the fine line metallization (in the x-y plane), used to implement the die-to-die (D2D)

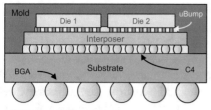

Differentiating Features:
- Multiple (2) die
- uBump Die Attach
- Si interposer

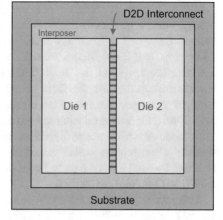

Fig. 2.2 Concept generic 2.5D SiP package—a cartoon illustration of a generic concept 2.5D SiP Package, showing cross-sectional and top-down views, and highlighting the major differentiating features (Si Interposer, Multiple Die, uBump attach), and identifying some standard package features (C4, BGA, Substrate, Mold)

interconnect. Interposer D2D Interconnect density is the parameter used to characterize a given 2.5D technology—typically of the order of um's width and space.

ii. **uBumps**: the level-1 interconnect (in z-direction) between the die and the interposer, usually implemented using some kind of solder bumps, Cu pillars, or direct Cu–Cu bonding, is also a differentiating feature. Typically, to leverage the high density x-y D2D interconnect, the level-1 bumps need to be on a correspondingly tight pitch—of the order of 10s of um.

iii. **Multiple Si die**: Si die may be manufactured in the same (homogeneous integration) or different (heterogeneous integration) Si technology, and may not necessarily be same size. In some instances, one or more of the die could also consist of 3D stacks of die.

Thus, 2.5D Technology is basically a packaging technology for housing 2, or more, die, side by side, in a single package, and enabling a fairly high density of die-to-die (D2D) interconnect.

2.2 2.5D SiP Technology Candidates

Unlike the case for traditional CMOS technology with its well-established dictates of Moore's Law and the ITRS-defined specific technology targets, there are no universal, industry standard, set of specifications and requirements for the 2.5D technology. In general, there is a requirement for fine line interconnect—but the exact specification of pitch, form factor, etc., is product specific. The interconnect density is a primary variable—analogous to the poly half pitch in CMOS technology—that is obviously driven by the nature of the D2D interface. For example, variants described in Sect. 2.3, including the Spilt Die architecture require 1000–2000 wires, and HBM Memory interface requires ∼2200 wires, etc. Depending on the die sizes involved this boils down to typically somewhere around 100–200 wires/mm die edge for the D2D interface (in addition to the usual interconnect required to support the I/O to the outside world, PDN delivery, and/or PoP memory). But individual products may require tighter (e.g., split die FPGA), or looser (e.g., modules with separate digital and analog chips), pitches. Similarly, there is no standard uBump pitch—other than that the current minimum is generally taken to be around ∼40u pitch—mostly based on the industry experience to date driven by the WideIO memory (Chap. 3).

This section describes the construction of the various candidate technologies that are most popular options for 2.5D Integration, mostly with a specific focus on the Split Die SiP implementation. Note that variations of the presented integration schemes are continuously developed and proposed, and that new solutions are expected to emerge. However, the construction schemes presented are representative of the current state of the art.

A summary of the 2.5D technology process flows is also presented. The intent is to illustrate the manufacturing flows at the concept level, identifying the major steps and features, rather than to discuss a specific process flow or detailed process recipes. The flows presented are generally representative, but any one specific implementation may vary, so that the specific sequence of steps, or distribution across the supply chain, may be different; presumably to optimize the cost or the operational logistics for a given product. Also included is a summary of the implications of a given SiP technology candidate on the Supply Chain—which is important for understanding the cost and risk associated with a given solution.

2.2.1 Through-Si Interposer (TSI)

Use of Si interposer inside a package is probably the most obvious way of implementing 2.5D SiP solution—and is hence the most common, and the oldest candidate. To date, all commercial products that leverage 2.5D Integration Technology, including Xilinx's Virtex (Xilinx 2016), AMD's Radeon-Pro (AMD 2016), and tsmc CoWoS-based products (tsmc 2016), are all variations based on Si Interposer technology. The principal reason is that it is perceived to be lowest risk 2.5D solution.

The intent of this type of packages is to leverage Si technology and foundry infrastructure to implement the fine pitch D2D interconnect, rather than to push the packaging technology to the lane/space specs beyond their current capabilities. The necessary features—including several dual-damascene Cu interconnect layers with <∼2u pitch, and Through-Si-Via (TSV—addressed in more detail in Chap. 3) typically with 10/100u diameter and depth respectively—are features that are not particularly challenging for Si class of technology. Most sources use equipment from ∼ 65 nm, or even older, generations of baseline CMOS fabs, to manufacture Si interposers. That is, *in general, the challenges with Si Interposer technologies— including the TSI and the LCIs type of solutions—are more to do with integration flow than with baseline interposer manufacturing.*

    i. ***TSI Construction***: TSI class of Si interposer technology has been used in first generation of 2.5D SiP products with large (> ∼ 200 sq mm pre-split Mono-Die SoC) target die sizes. TSI integration is typically perceived to be the *lowest risk* approach, as it leverages proven technologies from both Si and package domains, and is demonstrated in Low Volume Manufacturing (LVM) by the existing 2.5D SiP products (FPGA, HBM + GPU, CoWoS, etc.).

TSI implementation stacks active Si die on an interposer that includes TSVs, and then stacks the Si interposer on top of a conventional substrate, as illustrated in the sketch in Fig. 2.3. Concept SiP construction of GPU + HBM and Split Die FPGA, respectively, are shown. Note that the construction, and the numbers shown are meant to be illustrations of a concept and an order of magnitude, respectively, rather than representing a specific product

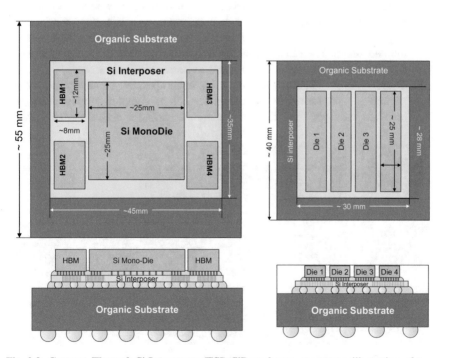

Fig. 2.3 Concept Through-Si-Interposer (TSI) SiP packages—a cartoon illustration of two concepts Through-Si-Interposer (TSI) SiP constructions, showing a cross-sectional and a top-down view of an Integrated GPU + HBM package, and a Split Die FPGA package, and identifying the typical order-of-magnitude dimensions

specification. Clearly these structures end up being large packages, several cms on a side, with a lot of Si content.

ii. **TSI Process Flow**: With this integration scheme a number (typically 3 +) of dual-damascene metallization layers and the blind Through-Si-Vias are defined in the Si Interposer wafers. The active Si Die are attached to the interposer (typically using tight pitch uBumps), the interposer is thinned to reveal the bottom of the TSVs (example process flow in Ch 3), and attached to the substrate (typically via C4 solder balls), which includes additional metallization layers to complete the routing and provide P/G planes, etc., and BGA balls. A concept process flow is illustrated in Fig. 2.4.

With this flow the interposer wafer is singulated after active die stacking, molding and interposer thinning and bumping, and the stacked interposer "die" is than attached to a substrate using C4 type solder balls. This is a version of the Chip-on-Wafer-on-Substrate (CoWoS) flow. The advantage of this approach is that the fine pitch uBump die-to-interposer attach of Si Die (and HBM cubes), leverages the excellent planarity of the full thickness interposer wafer, allowing use of cheaper mass reflow attach, rather than the die-by-die Thermal Compression Bonding (TCB). The disadvantage is that this is a "die-first" flow where the active Si die (the most expensive bits) have to go through the thinning and TSV reveal processing, as well as the interposer die-to-substrate attach and carrier de-bond processing. Whereas, in principle it is possible to define a die-last integration flow, where the interposers are thinned and attached to the substrate first, and the active Si die are attached to the interposer last, this does not seem to be practiced right now.

iii. **Supply Chain**: TSI flow, by definition, encompasses Si foundry and OSAT packaging and assembly technologies, with a complex supply chain that involves multiple foundries (for interposer and for active CMOS die), a

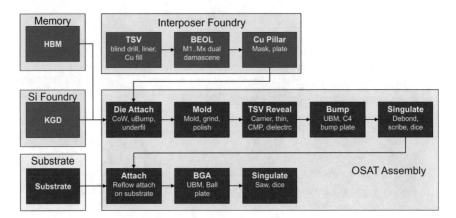

Fig. 2.4 Generic Through-Si-Interposer assembly flow—a concept high level assembly flow for Through-Si-interposer (TSI) Integrated GPU + HBM package, including the Si Interposer, Si KGD die, and Substrate components and highlighting the essential manufacturing and assembly steps

Substrate vendor (for manufacturing and testing the substrate), a memory supplier (in case of integrated HBM memory), and an OSAT (for integration and assembly). That is, the supply chain for implementing TSI structure is complex. Note that some foundries offer turnkey services, and will provide the interposer, as well as some of the post wafer fab process steps, or even the entire integration flow—typically in a bundle with the manufacturing of the active CMOS die. This "turnkey" option has its attractions and is often perceived as a low risk option, since the responsibility for yield is not distributed across multiple entities. This complex supply chain may not be readily compatible with the multi-sourcing constraints (requiring at least two sources for each essential component or process step) typically required by products that need rapid and reliable volume ramp.

Thus, the TSI integration scheme is quite expensive—since it includes both the interposer and the substrate, requires at least one incremental attach process step (relative to standard Mono-Die SoC packaging process), and one or more carrier insertions with temporary bond/de-bond steps, with potential yield loss at every incremental step, all accruing to a higher final product cost. Consequently, this integration scheme is applicable for very large active die areas—where the benefit is sufficient to cover the cost of the expensive package. Hence, this integration scheme is typically a candidate for the high end, low volume, high margin applications. TSI is not a realistic contended for high volume consumer applications, such as mobile phone; being too expensive, too large (in x-y plane), too thick (z-direction), incompatible with PoP implementation, etc.

2.2.2   Low-Cost Si Interposer (LCIs)

LCIs is a Si interposer based construction that targets lower cost applications (Kang and Yee 2013), potentially for use in mobile phones, and is hence built with smaller target split die sizes ($< \sim 10 \times 5$ mm) and smaller interposer footprint ($<15 \times 15$ mm). Note that the reduced target split die sizes, reduces the KGD benefit, and hence the budget for SiP package is also reduced (Sect. 2.3)

i. **LCIs Construction**: Typical concept construction is shown in the sketch in Fig. 2.5. Again, the structure and dimensions shown are for illustrative purposes, rather than being product specifications. For mobile applications, compatibility with PoP memory is also considered to be a requirement, along with a target total thickness (including the PoP memory) of ~ 1.0 mm, or less. In order to manage these cost and the z-dimension constraints, the TSI-like construction with substrate sandwich is not feasible, and all metal interconnects are therefore included in the Si interposer itself. That is, all wiring, including D2D wiring, I/O and PoP via routing, P/G distribution, TSV interconnect, UBM layers, and the BGAs are on the Si interposer. Low-Cost Interposer therefore

Fig. 2.5 Concept Low-Cost (Si) Interposer (LCIs) SiP package—a cartoon illustration of a Low-Cost Si Interposer (LCIs) SiP construction, showing a cross-sectional and a top-down view of a concept Split Die package, identifying the typical order-of-magnitude dimensions and showing a PoP'ed memory

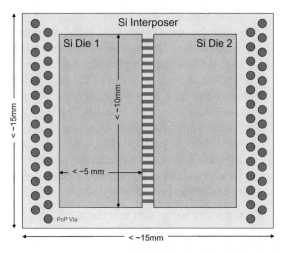

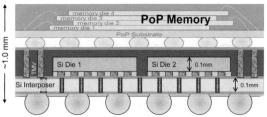

requires a few (2–3) layers of fine pitch interconnect on the top side (to address the routing requirements) leveraging foundry dual-damascene metallization process, TSV for vertical connection from top to bottom, and a layer or two of RDL on bottom side (for P/G distribution and UBM for BGAs), typically leveraging OSAT SAP process. In addition, PoP Vias need to be realized, along with PoP-side RDL, as necessary.

ii. *LCIs Process Flow*: concept LCIs process flow is presented in Fig. 2.6, below. As with the TSI flow, the fine pitch metal interconnect and the blind TSVs are defined on the topside of the interposer in a Si fab. The rest of the flow, including thinning of the interposer wafers to reveal the TSVs and uBumping—the so called "Middle End of Line" (MEOL)—can be done at either the Si foundry or at an OSAT. The choice of the MEOL source impacts the overall sequence of the integration flow (e.g., points in the flow where a carrier is attached), influences some of the choices of materials (e.g., PI versus SiN/SiO for backside dielectrics), and affects some process steps (e.g., Bumping of the interposer and "Temporary Bond/De-Bond" (TDDB) steps). Both Si Interposer flows (TSI and LCIs) shown here are for MEOL processing done at an OSAT—primarily because this appears to be the cheaper option that is more compatible with the kind of logistics used by products that compete in the high volume, consumer, markets—such as mobile phones—which tend to shape the mainstream technologies. The flow

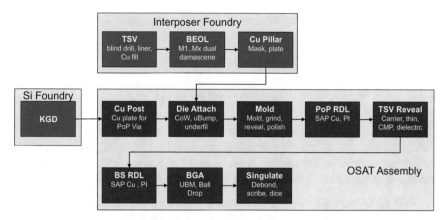

Fig. 2.6 Concept Low-Cost (Si) Interposer assembly flow—a generic high level assembly flow for Low-Cost Si Interposer (LCIs) package, including the Si Interposer and the Si KGD die components, and highlighting the essential manufacturing and assembly steps

presented also includes the incremental steps required to realize the PoP attach. In the case shown, the PoP Vias are formed by plating tall Cu posts on the interposer, followed by KGD attach, molding and PoP-Via reveal. Also shown is RDL on the PoP side, which may be required for the interface to the memory. Depending on the KGD thickness (and hence the height of the Cu posts), the plated Cu posts can be realized on a much tighter pitch than is currently possible with laser Through-Mold-Via (TMV) formation. That is, this feature is favored when spatial constraints do not allow use of conventional (cheaper) TMV PoP Vias. On the other hand, this flow requires carrier attach after PoP-side RDL formation, in order to thin the interposer wafer and reveal the TSVs. The flow shown is also die-first flow, exposing the KGD die to the entire assembly flow.

iii. ***LCIs Supply Chain***: LCIs supply chain is somewhat simpler than the one for TSI, but still, clearly requires a foundry source for the interposer processing and an OSAT for the MEOL and assembly. Note that the supply chain for Si interposers—especially for the low-cost kind—may be in jeopardy, since manufacturing Si interposers with only few layers of interconnect is not necessarily an attractive business proposition—even for an entity that has a fully depreciated Si fab line. Consequently, some of the turnkey services that bundle Si interposer with other manufacturing services offered by some foundries, may be a stable sourcing strategy. But it comes at a price, and may not be compatible with multi-sourcing requirements.

Thus, whereas this construction of Si Interposer is cheaper, and certainly thinner, than TSI, it is also quite complex, includes a number of challenging features, and is perceived to be a higher risk integration challenge than TSI. On the other hand, the fine line requirements associated with 2.5D integration, can readily be met. Furthermore, Si interposer has several other peripheral attributes—such as very good thermal properties (the interposer acts as a good heat spreader), excellent in-process planarity, etc.

2.2.3 Photo-Defined Organic Interposer (POI)

The basic idea with organic interposers is to add a few layers of high density interconnect to a standard organic substrate—thereby leveraging the existing normal supply chain and assembly flow, without the encumbrance of additional partners or processes (Kiyoshi et al. 2014; Ivankovic et al. 2015; Ryuta et al. 2015).

i. **POI Construction**: With this technology, most of the challenges (except die attach) of enabling high density D2D wiring are contained within the substrate manufacturing flow. Fundamentally this can be achieved by using semiconductor-type liquid photoresist litho processing to define one or two fine pitch layers on top of a standard substrate—instead of using a Si interposer. The standard substrate, including core + build up layers, then includes 2 to 4 additional metal layers of interconnect for general I/O routing and P/G distribution, as is normally used with Flip Chip BGA (FCBGA) packages. Usual UBM, solder mask, and BGA process complete the package. The resulting, and somewhat asymmetric, structure with all interconnect layers and vias contained in the substrate is illustrated in Fig. 2.7. Again, the numbers and construction in the figure are for illustrative purposes. In addition, for mobile applications, PoP Vias need to be realized, along with PoP-side RDL, as shown.

ii. **POI Process Flow**: the principal elements of the high-definition laminate Photo-defined Organic Interposer process flow are illustrated in Fig. 2.8 below. As shown, most of the flow is similar to the standard process used for manufacturing substrates for Flip Chip BGA packages. The additional high-definition layers required for the tight pitch D2D interconnect are defined using semiconductor-like photoresist process in combination with Cu sputtering of the seed layer and Cu electro-plating of the interconnect lines—as highlighted. Since substrates are typically manufactured in panel format, the photoresist process must use slit coating rather than the spin-on processing used in Si manufacturing. This approach is used for manufacturing displays and substrates, and as such is not unusual—albeit the fine line requirements for 2.5D integration do push the current state of the art of slit coating technology. Finally, standard FCBGA OSAT assembly process is used to attach the die, dispense the underfill and mold the packages. Again, the tight ∼40 um uBump pitch die attach pushes the standard substrate process state of the art, and additional precautions may be required to ensure adequate planarity during the attach—such as use of thicker die, and or thermo-compression bonding process. The process illustrated assumes that the spatial constraints allow the use of standard TMV processing to define the PoP vias, but additional PoP-side RDL routing is also included—as may be necessary. If the spatial constraints (die and package size) do not allow

**Fig. 2.7 Concept
Photo-defined
Organic-Interposer
(POI) SiP package**—a
cartoon illustration of a
Photo-defined Organic
Interposer (POI) SiP
construction, showing a
cross-sectional and a
top-down view of a concept
Split Die package, identifying
the typical
order-of-magnitude
dimensions and showing a
PoP'ed memory

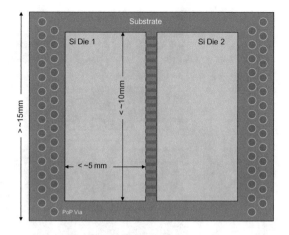

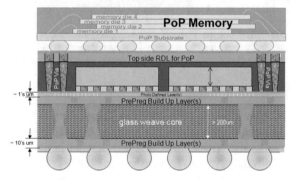

for large TMV PoP vias, then a tall Cu post plating process—similar as described for LCIs—can be used.

That is, whereas the basic process is like the standard FCBGA substrate flow, the typical requirements associated with the 2.5D integration do pose incremental and significant new challenges.

iii. *POI Supply Chain*

One of the key attractions of this integration technology is that it does not require incremental suppliers in a supply chain, and most of the logistics is much like the standard FCBGA flow. From the operations point of view, it is therefore perceived as a lower risk implementation. On the other hand, the fine line requirements do push the current substrate state of the art, and the substrate manufacturers may need to upgrade their entire infrastructure—such as the clean room standards, inspection and test equipment, procedures, etc.—in order to enable high volume manufacturing of $\sim$2 um line/space interconnect needed for 2.5D SiP packages.

That is, POI integration scheme is an evolutionary approach that calls for some optimization in order to be implemented in HVM. The basic technology has

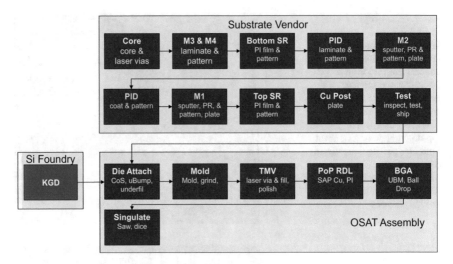

Fig. 2.8 Concept Photo-defined-Organic-Interposer assembly flow—a generic high level assembly flow for Photo-defined Organic Interposer (POI) package, including the Substrate and Si KGD die components and highlighting the essential manufacturing and assembly steps

multiple operational and technical advantages. Notably, there is a good proof-of-existence for managing issues caused by the mismatch of Coefficient of Thermal Expansion (CTE) between Si and PCB, such as Board Level Reliability (BLR) challenges—something that is an issue for LCIs implementation. Hence, POI offers a promising path for scalability to larger form factors (X-Y dimensions) SiP products. On the other hand, the dimensional stability of the materials used in substrate construction as well as some of the process techniques, are challenged by the fine dimensions required. Issues such as adhesion of the spin-on layers and planarity for uBump attach have to be managed within the target constraints (e.g., 2 u line/space, 40 u uBump pitch, 1 mm overall thickness, etc.). Note also that managing Signal Integrity issues is a bit more challenging with POI technology, due to use of relatively thick dielectric layers—which reduces the effectiveness of ground planes (Sect. 2.4).

2.2.4 Low-Cost Glass Interposer (LCIg)

The concept of using glass to replace organic laminate substrate in generic microelectronic packaging has been around for a while (Sukumaran et al. 2012; McCann et al. 2014; Huang et al. 2014; Chen et al. 2014)—driven by its good mechanical and electrical properties, and attractive cost structure. Since the core material in regular substrates is a glass fiber weave embedded in resin, current standard substrates already include a certain content of glass. However, the tighter

Fig. 2.9 Concept Low-Cost (Glass) Interposer (LCIg) SiP package—a cartoon illustration of two concept Low-Cost Glass Interposer (LCIg) SiP constructions, showing a cross-sectional and a top-down view of a Split Die Glass Core Substrate package (**a**), and a Glass Body package (**b**), identifying the generic order-of-magnitude dimensions and showing a PoP'ed memory

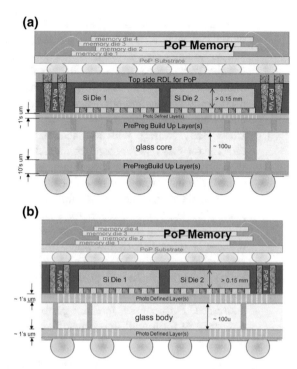

line/space specs required for 2.5D Integration, pushes the need for increased dimensional stability, and hence naturally favors higher glass content.

Glass is a clean, uniform, flat, material whose mechanical properties (e.g., CTE) can be tailored within reason to specific applications, and with superior electrical insulation properties. In principle, it could be made at very low cost, potentially using roll-to-roll manufacturing. It is ubiquitous in the display industry (phones, PCs, TVs, etc.), and has a very good track record as the foundation material with panel sizes now reaching beyond Gen 10 ($\sim 2.8 \times 3.0$ m) sizes. However, some of the materials (e.g., Cu conductors) and dimensional (e.g., fine line/space specs) necessary for implementation in 2.5D packages are different.

i. ***LCIg Construction***: Two basic approaches have been proposed, albeit neither has been demonstrated in full for multi-die SiP constructions:

   (a) Glass Core Substrate (Panel): with this approach glass sheets, rather than glass fiber weave embedded in resin, are used to form the core of the substrate, presumably resulting in a package with superior dimensional stability, and hence more compatible with fine line requirements. The integration scheme is illustrated in Fig. 2.9a showing the concept construction and target dimensions. The change from the flow currently used to manufacture package substrate panels is relatively minor, since the pre-preg buildup layers (or similar alternatives) and Cu RDL technology could be used. The new required processes include formation (drill and fill) of Through-Glass-Vias (TGV), managing adhesion of buildup layers to (smooth) glass, and managing the handling of (brittle) glass sheets. In addition, for fine line capability, incremental modules—perhaps such as used in the POI technology—are required to realize $\sim 2u$ line/space specs and $\sim 40u$ uBump attach.

(b) Glass Body (wafer): with this scheme, metallization runs are deposited directly on glass wafers, and the standard substrate manufacturing flow with organic buildup layers is eliminated entirely. Analogous construction is used for manufacturing modules for integration of passive components for rf applications—leveraging the superior insulation characteristics of glass. The target construction for 2.5D SiP applications is illustrated in Fig. 2.9b, showing concept structure and dimensions. The manufacturing process is more like Si fab flow, based on glass wafers, with standard spin-on photoresist, CVD, and etch modules, than like substrate process with slit coating resist and plating processes. However, the TGV module needs development and metal-to-glass adhesion challenges need to be addressed. Furthermore, technology for deposition of insulators on glass necessary to form multi-level metallization is also required and still to be developed.

ii. ***LCIg Process Flows***: the concept process flows for the two types of the LCIg constructs are illustrated in Fig. 2.10 below. Note that whereas elements of both flows have been demonstrated in labs on engineering bases, neither of the flows have been proven in full, with the target specs for SiP implementation. Hence, the figure illustrates the fundamental concept and principal steps required, rather than a specific implementation.

As shown, the manufacturing of glass substrates—either in panel or in wafer format—is performed by a separate manufacturer—such as (Corning 2016) or Asahi (AGC 2016). Since formation of TGVs, and especially the TGVs with a diameter of the order of ~ 10 um (required for digital SIP applications), requires specialized equipment, it is most likely that the glass vendors will develop the process for drilling the vias, and potentially, even for filling them with Cu. For implementation of panel level Glass Core Substrates for 2.5D SiP applications, this would then be combined with standard build up process and a POI-like processing to realize the necessary fine pitch interconnect. For implementation of wafer-level Glass Substrates, this would be combined with Si Fab or LCD display-like processing to realize the interconnect. Note that, as shown, with ~ 100 u thick wafers, it is likely that incremental carrier bond/de-bond steps would be needed in the flow. Finally, standard OSAT assembly flow with mold, TMV PoP vias, and BGA attach is shown. As discussed above, tight spatial constraints may force use of process other than laser TMV PoP via.

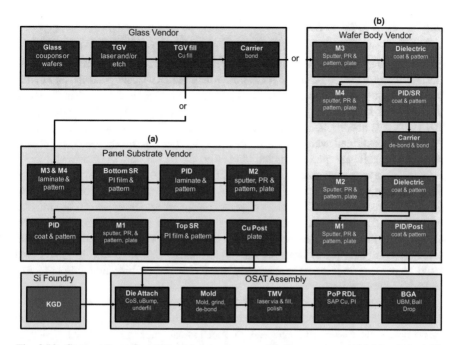

Fig. 2.10 Concept Low-Cost (Glass) interposer assembly flow—a generic high level assembly flow for Low-Cost Glass Interposer (LCIg) package, including the Glass and Si die components and highlighting the essential manufacturing and assembly steps both Glass Core Substrate and Glass Body implementations

iii. ***LCIg Supply Chain***: manufacturing of LCIg would involve at least one more class of vendors in the supply chain—the glass suppliers. Clearly their technology and the accompanying design rules would need to be integrated into the overall flow, and the design rules harmonized across glass vendor, substrate vendor, and OSAT technologies. This is one of the reasons that this option is perceived as being a high-risk solution for 2.5D SiP implementation

Whereas glass interposer, in either construction scheme, is attractive due to the desirable material characteristics relative to the substrate technology, none of the current applications that leverage glass properties quite match the requirements for 2.5D SiP implementation. Technology for building displays uses different interconnect materials and line width specifications. Technology for rf passive components favors thicker glass wafers, bigger TGVs and coarser/thicker metallization layers. Thus glass-based interposer construction for digital SiP application is perceived as being a relatively high risk, disruptive opportunity. In addition, LCIg technology may encounter other peripheral challenges—such as managing thermal and thermo-mechanical characteristics, CTE mismatch versus PCB, Si, etc.

2.2.5 Fan Out WLP

Fan Out Wafer-Level Packaging (FO WLP) technology is a variant of Wafer-Level Ball Grid Array (WLB) class of packages, where the BGA balls, and the associated interconnect are deposited directly on the Si die—without an independent substrate or interposer layer (Tseng et al. 2016; StatsChipPac 2016; Yu 2015; Yoon et al. 2013). This is accomplished by placing KGD die on a carrier and molding this to form a "reconstituted wafer"—which can then be handled through various process steps to deposit and pattern metal (typically Cu RDL) and dielectric (typically Polyimide) layers, necessary to route the I/Os, distribute P/G and carry the BGAs. With the original WLB packages, the BGA balls were confined to the footprint of the die, whereas with Fan Out class of packages the interconnect and BGAs may go beyond the die footprint. Consequently, this class of packages are intrinsically cheap, since there is no interposer that has to be paid for, and are intrinsically thin, since there are only a few layers of RDL interconnect that are added to the die and mold thickness; making it particularly attractive for mobile and perhaps, IoT wearables applications.

On the other hand, FO packages are "floppy" single sided structures—without a stiffener of any kind, and using materials with highly asymmetric mechanical properties. This makes management of mechanical warpage challenging—especially for larger form factors calling for finer interconnect features (Sect. 4.3)

i. **FOWLP Construction**: Wafer-Level Fan Out Technology typically results in a stack up roughly as illustrated in Fig. 2.11. The configuration shown includes PoP vias—in this case based on "eBar" or "pcBar" approach with an organic substrate interposer in case PoP RDL is required. PoP Vias can also be implemented using tall plated Cu posts—analogous to the ones described for LCIs. FOWLP construction ends up with very thin packages—since there is no substrate and/or interposer thickness to account for-easily meeting thicknesses requirements of the order of ~ 1 mm or less. Overall thickness is dictated by the thickness of the Si die, the overmold, and, in case of PoP implementation, the organic interposer, and is constrained by mechanical integrity specifications (warpage, planarity, etc.). The other key figure of merit for this construction is "Fan Out Ratio," i.e., the ratio of the area of the die to area of the overall package. FO Ratio is a metric that is important for the mechanical characteristics of the construction, affecting in-process and end-of-line warpage and planarity. Managing these characteristics is a critical issue for this type of package and impacts the cost and the overall process flow. Note that the challenge is different for high versus low fan out ratio packages. Fundamentally High FO Ratio packages have to balance the CTEs of region that is dominated by mold (around the package periphery) and region that has both mold and Si (central region), whereas Low FO ratio packages are dominated purely by Mold and Si characteristics. Note that whereas the construction illustrated is a 2.5D SiP implementation, that requires the fine pitch D2D interconnect, this class of

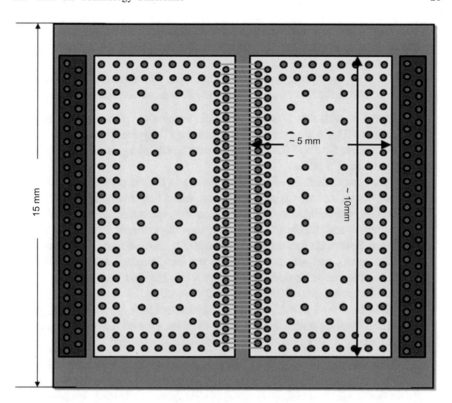

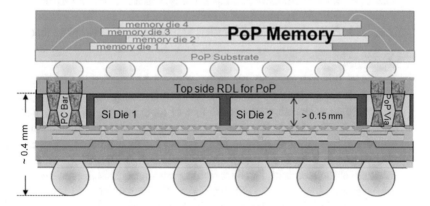

Fig. 2.11 Concept Fan Out Wafer-Level Package (FOWLP) SiP package—a cartoon illustration of a Fan Out Wafer-Level-Package (FOWLP) SiP construction, showing a cross-sectional and a top-down view of a concept Split Die package, identifying the typical order-of-magnitude dimensions and showing a PoP'ed memory

packages is getting considerable traction with mono-die SoC products—due to the thickness and cost benefits—in which case fine pitch (~ 4 um) routing may not be required.

ii. **FOWLP Process Flows**: there are several types of implementation flows typically practiced for this class of packages. These can be characterized as

   (a) Face Up: with this flow the KGD die include Cu pillars on the chip I/O pads. KGD die are placed on a wafer-shaped (typically glass) carrier with their active face up. The structure is then coated and ground and polished to reveal the Cu posts.

   (b) Face Down: with this flow KGD die are processed up to the top I/O pads and do not include Cu pillars. Die are placed face down on a wafer-shaped carrier, and molded to form the reconstituted wafer. The carrier is then removed, and the die I/O pads are opened by standard lithography process

   (c) Mold First: with this approach the wafer is molded first (either face up or face down), and acts as a carrier for the interconnect that is defined on the reconstituted wafer starting with fine pitch M1 first, and building up M2 and M3 layers and BGAs sequentially

   (d) Mold-Last: with this approach the interconnect RDL layers are formed on a carrier, starting with M3 and ending the stack up with fine pitch M1, the die are attached only then, and the wafer is finally molded

The flow illustrated in Fig. 2.12 below is a generic flow, with both, the pcBar and Cu Post options for PoP via shown. The various variants of the flow tend to affect the carrier insertion points and the associated bond/de-bond processing. The common denominator among all the variants is the RDL interconnect processing—typically based on standard Photoresist and SAP-plated Cu process, with Polyimide (PID) dielectrics, and ball drop BGAs. Note that in spite of the apparent simplicity of the construction—this flow has a number of challenges.

Realizing the flatness and planarity of the reconstituted wafer required in general, but especially for fine pitch interconnect, may be challenging. This is because the construction uses a number of materials (Si, PID, Cu) with very different CTEs. Balancing the various stresses at high temperatures, for in-process control, and at room temperature, for final product specification, is challenging in a package that does not include a "stiffener"—such as a thick substrate associated with FCBGA class of packages. This is exacerbated when the package is large and the FO Ratio is small—thereby putting a practical limit on package size that can be realized in this technology.

iii. **FOWLP Supply Chain**: the supply chain for FOWLP packaging is relatively simple, since most, if not all of the processing is performed either at the foundry (Si KGD, Cu pillars) or OSAT (all the rest). The exception is when the pcBar approach is used for implementing PoP Vias, where the pcBars and the PoP Substrate are typically provided by the third-party substrate vendors—albeit neither of the subcomponents are complex or expensive. Note that some

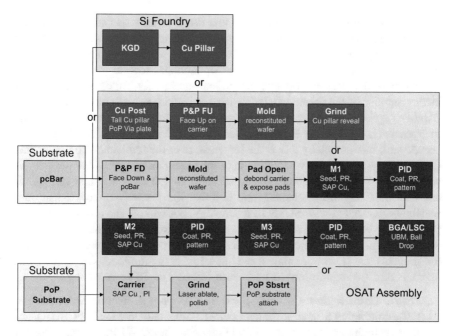

Fig. 2.12 Concept fan out WLP assembly flow—a generic high level assembly flow for Fan Out Wafer-Level Package (FOWLP) package, including the Si KGD die and PoP Substrate components and highlighting the essential manufacturing and assembly steps, with options for both plated Cu pillar + Face Up and PcBar + Face Down implementations

foundries offer WLP—and recently even FO WLP class of processing—as an extension of their Si fab services, potentially making the supply chain a "one stop shop" case.

FO WLP technology is very attractive especially for mobile market due to the intrinsic thickness and cost structure. On the other hand, FOWLP packages have intrinsic limitations in terms of the RDL pitch and package footprint, and as such may not be easily extendible to potential future 2.5D SiP requirements. Similarly, the supply chain is simple, but the technology, at this time, may not be readily multi-sourced, since it is not (yet) ubiquitous and since some of the flow variants may not be readily interchangeable. Finally managing mechanical integrity, and potentially Signal Integrity issues with FOWLP technology may be challenging due to its floppy structure, and the use of thick PID layers (like POI), respectively.

2.2.6 Hybrid Technologies (SLIT, SLIM)

Hybrid Technology is a term applied here to an emerging class of integration technologies that combine some of the 2.5D technologies. Specifically, there are several "TSV-Less" technology options that use Si Interposers to realize the features that are beyond the scope of normal packaging technologies, and then use normal packaging flow for everything else—thereby presumably obtaining the best of all worlds. That is, pushing packaging technologies toward lines finer than 2–5 um line width/space is possible, but difficult and costly, fundamentally due the incompatibility of the organic materials used in packaging with of the dimensional stability required for higher density interconnects. On the other hand, standard Si Interposers (LCIs, TSI) are expensive, and use of TSVs for high speed interconnect is awkward, fundamentally due to the lossy nature of the semiconducting materials. Consequently, an alternative solution is to integrate portions of Si Interposers, but without the TSVs, with either organic substrate or FO flow, by removing the Si substrate bulk. Hybrid implementation like that has the interconnect density and scalability associated with Si BEOL technology, but without the cost structure and performance limitations associated with manufacturing and assembly of complete Si Interposers. These hybrid technologies are hence potentially attractive for 2.5D SiP products that (a) require high-speed interconnect (no limitations associated with TSV), and/or (b) require low-cost structure (no cost of TSV processing).

i. **Hybrid Construction**: Currently there are two types of hybrid constructions—those combining Si Interposer with Organic Substrate flow (as in FCBGA), and those combining Si Interposer with FO flow (as in FO WLP). In either case the construction ends up looking like a substrate SiP or FO SiP, respectively, but with a very thin layer of Si BEOL interconnect sandwiched between the active Si die and the package:

(a) *Hybrid TSV-Less Si Interposer + Substrate* Construction, notably pioneered by SPIL and Xilinx (Kwon et al. 2014) and named Silicon-Less Interconnect Technology (SLIT). The construction illustrated in Fig. 2.13b uses a TSV-less Si Interposer which is then ground away, leaving just the several layers of BEOL interconnect layers, stacked on top of a conventional 8–14 layer substrate. The primary motivation for eliminating the Si interposer for this type of application—large Split Die FPGA—is believed to be performance, i.e., TSV in Si Interposer—a semiconductor—is a lossy feature not compatible with very high speed interconnect ($< \sim 20$ GHz/sec). The SLIT construction eliminates the TSVs altogether

(b) *Hybrid TSV-Less Si Interposer + FO* Construction, notably pioneered by Qualcomm and AMKOR (Kelly et al. 2015; Kelly et al. 2016) and named Silicon-Less Integrated Module (SLIM). The construction illustrated in Fig. 2.13a also uses a TSV-less Si Interposer which is then ground away, leaving just a layer of BEOL interconnect, which is then integrated with 2–3 standard FO RDL layers. The combination of Si Interposer interconnect density advantages with the Fan Out form factor and cost advantages, makes this option interesting for mobile and IoT/Wearables markets. SLIM is fundamentally a FOWLP package with a single fine pitch dual-damascene interconnect layer. The motivation here was more to eliminate the cost structure of Si Interposer, while preserving the high

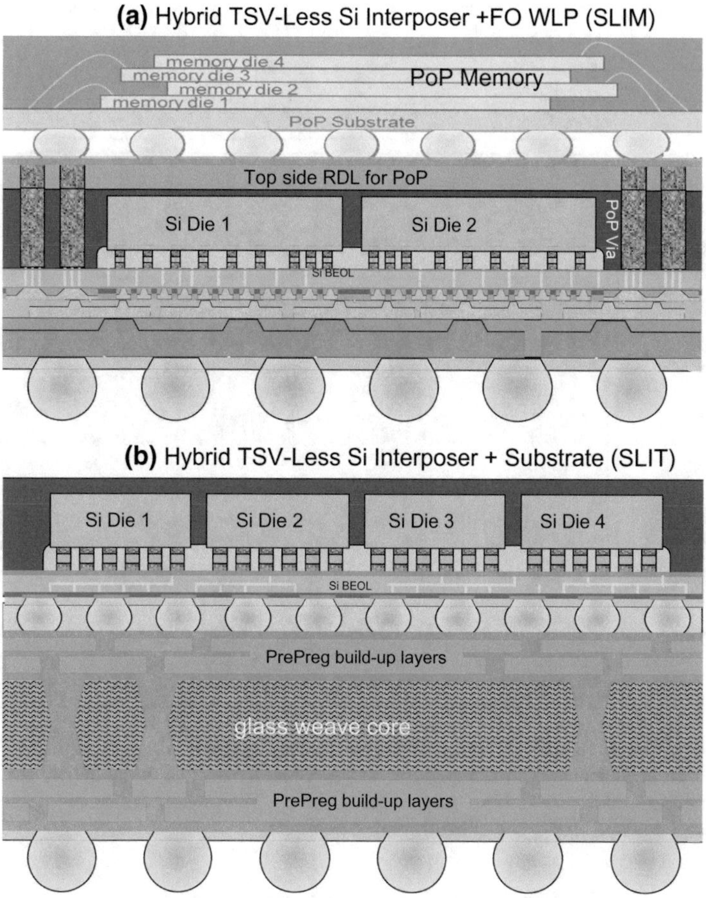

Fig. 2.13 Concept hybrid TSV-Less Interposer SiP packages—a cartoon illustration of two concept Hybrid SiP constructions, showing a cross-sectional view of a TSV-Less Si Interposer + Substrate (SLIT) package, and a TSV-Less Si Interposer + FOWLP (SLIM) package

density interconnect (Si technology) as well as thickness and cost (FO technology).

ii. *Hybrid Process Flow*: Fig. 2.14 below illustrates the conceptual process flow for manufacturing of either type of Hybrid packages. The SLIM flow (Si Interposer + FO) is shown with the PoP option, based on Cu pillar, whereas the SLIT (Si Interposer + Substrate) flow excludes the PoP altogether—due to the current target application space for the two technologies

As shown, most of the SLIM flow is standard Fan Out flow, performed at an OSAT (presumably at the cost and margin associated with that sector of the industry), but the starting material is a Si wafer with one layer of metal

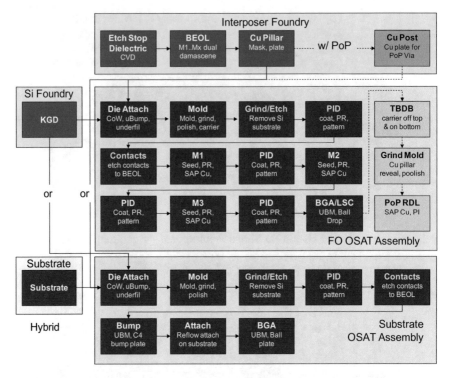

Fig. 2.14 Concept hybrid TSV-Less Interposer assembly flow—a generic high level assembly flow for Hybrid packages, including the major components and highlighting the essential manufacturing and assembly steps for both, Interposer + Substrate (SLIT), and Interposer + FOWLP (HDFO or SLIM) implementations

interconnect (easily) manufactured at a foundry, with the integration requiring only a few nonorthodox steps.

The steps associated with the key differentiating feature of the hybrid packages —removal of Si Interposer substrate and opening of the contacts to the back of the BEOL stack—are highlighted. The flow shown for the Si Interposer + FO version is the mold-first (die-first) option. A mold-last (die-last) flow can also be defined, but it would require different carrier insertion points. However, die-last flow does not have the cost penalty associated with the yield loss of good die attached to a bad package. Similarly, the PoP flow shown is based on plated Cu pillar, but as is the case with other similar packages, a TMV approach could be used if the spatial constraints allow the correspondingly coarser pitch. Note that since all Si is eventually removed, the starting material for the Si Interposer does not have to be of high semiconductor grade, thereby providing considerable cost savings.

The principle risks with either of the Hybrid Package manufacturing flows is managing the mechanical integrity and planarity during the processing— especially during the critical steps of Si substrate removal. Note that, as shown, a dielectric layer must be included at the bottom of the BEOL stack during Si Interposer manufacturing, in order to provide an etch-stop layer with sufficiently different selectivity versus Si, to allow for clean and complete removal of the Si substrate. Managing that step, and the following steps required to integrate the BEOL stack with the standard package (either substrate or FO) is critical.

iii. *Hybrid Supply Chain*: The supply chain for building the SLIT and SLIM Hybrid packages is similar to the standard FCBGA and FOWLP, respectively, with the additional requirement for managing the Si Interposer source and Si Substrate removal. This imposes an incremental requirement for specialized equipment and process modules which are not typical for standard OSAT flows. Furthermore, in addition to the risks associated with the supply of standard Si Interposers mentioned above, the incremental processing required to provide the etch-stop dielectric at the bottom of the BEOL stack, drives incremental equipment/process requirements that further jeopardizes the supply of Si Interposers for Hybrid packages. On the other hand, the main attraction of the hybrid type of a package is that it does not push any supplier (OSATs, Substrate Vendors, Si foundries) beyond their natural comfort zone, and hence does not require much of an incremental investment in manufacturing infrastructure.

Hybrid Packaging technology option is very attractive, not only due to the supply chain considerations mentioned, but also because it is naturally very scalable to even finer line/space specs than is currently required—but without placing incremental strain on the supply chain. This makes this class of packages potentially very attractive for the SiP applications targeting not only the Split Die but also the Small Die Value Proposition (Sect. 2.3). This may give hybrid technology traction in the mobile and IoT Wearables sector—markets that stress power, form factor and cost over other attributes. Hybrid packages excel in exactly these attributes.

2.3   2.5D SiP Technology Value Propositions

There are multiple different system partitions and integration schemes which could leverage 2.5D Technology to generate incremental Value Propositions. In general, 2.5D Technology value propositions can be segregated into two groups, and are:

At the System Level: (versus standard implementation using multiple IC components)

i. Better *performance* than possible with multiple single chip components: by placing multiple ICs closer together in a single package, the I/O interconnect loads are reduced resulting in better performance and/or lower power

ii. Better System *form factor* than possible with multiple single chip components: by placing multiple die in a single package the system form factor can be reduced by eliminating some packages—and their footprint from the PCB

iii. Better *cost* than possible with multiple single chip components: by placing multiple die together in a single package better system cost could potentially be achieved by eliminating the cost of one or more packages

At the Component Level: (versus standard implementation using a mono-die SoC)

iv. Better *Technology Utilization* than Mono-Die SoC: by breaking up a monolithic SoC into multiple die in a single package, each die can be implemented in its optimal Si technology—e.g., analog versus digital subcomponents could be implemented in suitable technologies, producing easier design and potentially better power-performance tradeoff

v. Better *Scalability* than Mono-Die SoC: splitting a mono-die SoC into several die allows reuse of IP housed on one of the die which may not require More-Moore type of CMOS scaling, and thus having to redesign less new IP that is housed on the other die which does need to scale with Moore's Law.

vi. Better *Cost* than a (Large) Mono-Die SoC: for applications, such as high-end servers or FPGAs, performance is driven by content consisting of many, parallel functions, so that high-end implementations require very large die. In the extreme, the die size may be butting against mask reticle sizes. In situations like that splitting a large SoC into several smaller die may be more cost effective versus for example reticle stitching

vii. Better *Cost* than (Medium Size) Mono-Die SoC: for applications, such as the Processors for PCs, tablets, and phones, die sizes may be growing beyond the "sweet spot" of a given technology node, resulting in poor yields. In situations like that splitting a SoC into several smaller die may be more cost effective since yield of several smaller die may be better than yield of a single larger die.

Note that the value propositions are segregated into two groups; namely those that are leveraged at the system level, and those that are leveraged at the component level. As illustrated in the sketch in Fig. 2.15, with the first group, the value proposition would be typically weighed versus "standard" implementation using multiple IC components on a PCB, and as such, the value proposition could be monetized at the system level—e.g,. mobile phone that is smaller, thinner, better battery life, etc. With the second group, the 2.5D SiP Value Proposition is typically weighed against a comparable mono-die SoC implementation, and as such the value proposition could be monetized at the component level—e.g., lower cost, smaller package, better IC power-performance...

Thus, as shown, SiP implementation leveraging 2.5D Technology has two principal Value Propositions, namely the "Integration" value proposition at the system level, and "Split Die" value proposition at the component level.

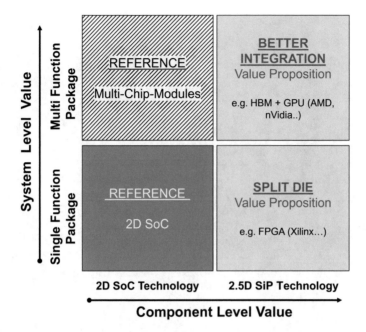

Fig. 2.15 2.5D integration technology value propositions—an illustration of the types of value propositions that are derived from 2.5D integration technologies, and highlighting the combinations that produce Component Level and System Level differentiation versus 2D SoC and MCM packages respectively

2.3.1 System Integration Value Proposition

Typical representative SiP implementation scheme that demonstrates the 'Better Integration' value proposition, at its best, is the integration of high-end GPU processor with the High Bandwidth Memory (HBM), as implemented for example by AMD (Black 2012; Lee and Black et al. 2016).

The principal value proposition is superior power-performance, provided by the High Bandwidth Memory, and Memory to Logic interconnect. This memory architecture provides higher bandwidth (128–256 GB/s) at superior power efficiency (35 GB/s per Watt) versus the conventional GDDR5 memory (28 GB/s with ~10 GB/s per Watt), by using a wide interface (x1024 versus x32, respectively). 2.5D Technology is an enabler that releases this value proposition by providing an interconnect solution for wide memory interface to the GPU, through use of a Si interposer, as shown in Fig. 2.16. The full value proposition is realized at the system or card level, even though the cost of this integrated SiP module is higher than the combined cost of the individual components. This SiP module has been announced by AMD in its Radeon™ R9 Fury X products, in May 2015, and is currently commercially available. The panels provided by AMD as a part of its

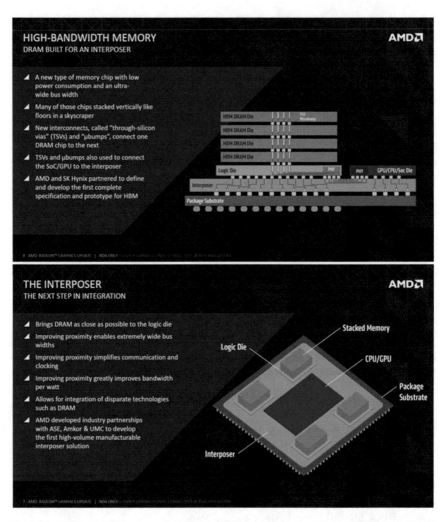

Fig. 2.16 Integrated SiP value proposition—AMD presentation panels illustrating the 2.5D SiP solution with its HBM memory + GPU die implementation used in Radeon product family, along with the stated value propositions (AMD 2016)

product announcement (see amd.com), is probably as good a way of presenting the value propositions as there is, as shown in Fig. 2.16.

Other products using similar architectures, with 2.5D Integration of wide HBM memory and a processor die on a Si Interposer for High Performance Computing, Networking, and Graphics applications have also been discussed in the industry (eSilicon 2016; Remi 2012; Kim and Kim 2014), and are either commercialized or are on verge of commercialization. Most of these SiP products target high-end applications where performance, and possibly power efficiency, trump cost considerations. Whereas this tradeoff is clearly currently acceptable for low to medium volume applications associated with the high-end computing, cost typically trumps raw performance for the high volume consumer applications, such as mobile phones, tablets and PCs.

For the purposes of this book, this specific class of implementations of 2.5D integration technologies is not discussed in detail—mostly because they are already in commercial domain, as implemented in announced products. However, many of the tradeoffs and considerations discussed further in the book do apply equally to the system level "Integration" value proposition as they do to the component level "Split Die" value proposition. Hence, the rest of this chapter is mostly based on the "Split Die" value proposition—primarily because this is more applicable for the high volume, consumer, applications which ultimately will shape the More-than-Moore technology mainstream, and because its technical tradeoffs tend to encompass both value propositions.

2.3.2 Split Die SiP Value Proposition

This section discusses the value proposition of a Split-Die implementation—i.e., breaking up a Mono-Die SoC into multiple die that are then reintegrated in a single SiP package. Note that the focus here is on *cost* benefits, i.e., Average Unit Cost (AUC) in High Volume Manufacturing (HVM). Since the principal variable used to gauge adoption of a technology candidate for consumer applications is cost. Other metrics (power, speed, form factor..) are basic specifications that have to be met. AUC is the primary metric that determines the viability of a disruptive 2.5D Integration technology for high volume applications.

The source of the potential AUC benefits of a Split-Die solution is the exponential nature of the Si Yield Curve, i.e., Yield is related to Die Area (A) via a process specific constant (Defect Density D_0), as described by, for example, Poisson Yield Model: $Y = \exp(-D_0 A)$.

More complex models (Bose-Einstein, Stapper, Murphy...) add factors that account for design and/or process complexity, and different defect distributions, but for a given technology and design type, the basic Poisson exponential relationship is a reasonably good model. Thus, for a given process technology, the pure, fundamental Si cost—derived from first principles—of two die with half an area (A/2) each, is better than the cost of a single die of that given area A. This cost advantage of a symmetrical die split (into two die with equal A/2 area) is then a function of the original Mono-Die die Area (A) and process maturity, described by Defect Density (D0) constant, as illustrated in Fig. 2.17. "Cost Advantage" here is the cost of two half-area Split-Die relative to the cost of a full area Mono-Die, expressed as % of Mono-Die cost of, i.e., % over and above a simple linear relationship between area and cost). Thus, as shown, splitting a small die (e.g., $< \sim 50$ mm^2) results in relatively small cost advantage ($< \sim 10\%$)—for a reasonably mature process. But splitting a large die (e.g., $> \sim 200$ mm^2) results in significant cost advantage (e.g., $> \sim 25\%$), especially with immature technologies characterized by relatively

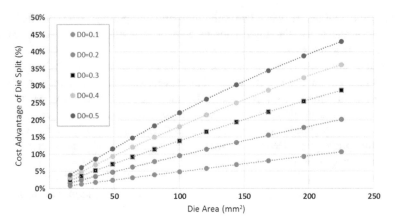

Fig. 2.17 Split Die value proposition—a plot of die level (KGD) cost advantage of a symmetrical Split Die (two die with equal A/2 area) relative to the cost of a Mono-Die die SoC (Area A) versus the area of the Mono-Die SoC, with process maturity, defined by Defect Density (D0), as a parameter

large D0 ($>\sim 0.3$ def/cm^2). As the process technology matures toward its entitlement D0 (~ 0.2 def/cm^2), Split-Die benefits larger than $\sim 10\%$ can be realized only with die sizes A $\gg \sim 100$ mm^2.

Consequently, this type of a Value Proposition is an *"Area Based"* Value Proposition—to differentiate it versus other potential Split-Die value propositions that have different area dependence (Sect. 2.3.3).

Note that the numbers illustrated here are approximate, and depend on many second-order variables—but the principals and the trends are valid.

Note also that in case of server, memory, and/or FPGA die which may use various redundancy and/or resilient architecture schemes, such as availability of spare features which can be fused in to repair a failing die, the Split Die area-based value proposition is substantially changed even with very large die, i.e., the basic analyses here is substantially changed by the architecture of the redundancy schemes.

However, Split-Die SiP implementation can also lead to cost increases—typically due to more expensive 2.5D packaging technology, or higher cost of test, or the need for different interfaces, etc. Clearly the goal is to leverage the benefits (Split-Die KGD advantage) to cover all potential increases in costs, so that the final AUC is reduced, i.e., some net savings are left over to drive a better margin.

Thus, for a given D0 value, there is a Mono-Die size that results in a break-even point where the Split Die KGD benefits are exactly equal to increased SiP package costs. Below that die area, the use of Mono-Die SoC is actually more cost efficient than use of Split-Die SiP. This is illustrated in the Fig. 2.18, plotting the locus of the Mono-Die die size that results in break-even cost as a function of D0, with cost of the interposer as a parameter. For a given curve, the area above the line represents region where Split-Die is favorable, whereas the area below the line represents situation where Mono-Die is a lower cost solution. There is clearly a number of

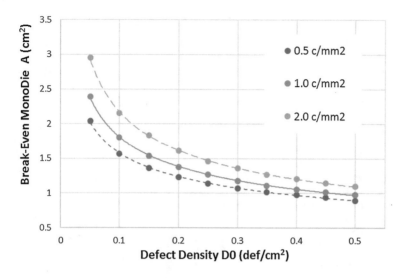

Fig. 2.18 Mono-Die SoC versus Split-Die SiP Break-Even Locus versus defect density—a plot of break-even locus of Mono-Die SoC area where the cost of Mono-Die SoC is equal to the cost of Spilt Die SiP versus Process Defect Density, with the cost of the interposer as a parameter, and with typical assumed values for other variables

variables that have to be constrained to create such a plot (wafer price, interposer/package and assembly cost, D2D bump pitch, number of D2D wires, die aspect ratio, area ratio, etc.), but the plot does illustrate a key point: **Split Die Value Proposition favors Big Die and Immature Processes**.

2.3.3   Small Die SiP Value Propositions

A key attribute of the 2.5D technology is that it pushes interconnect density. In case of the above Value Propositions (Integration and Split Die), this is required to enable a parallel interface between multiple die co-located in a single package. However, high density interconnect is also necessary to support a standard interface to a small die; not so much because the pin count is increased, but because the die periphery is decreased. The Value Propositions described below are enabled by 2.5D interconnect technology due to their use of small die and are not just straight area-based arguments.

i. *Scaled Split Die SiP Value Proposition*: As discussed above (Sect. 2.3.2), the Split-Die value proposition is related to the area, i.e., splitting a large Mono-Die SoC brings larger benefits than splitting a smaller Mono-Die SoC. Note that the value proposition is always there, i.e., two smaller die will always yield better than one large one—it is just that the KGD benefits for small die are lower than corresponding SiP costs. However, with the Si cost/unit area of advanced CMOS nodes going up, and the cost of the 2.5D

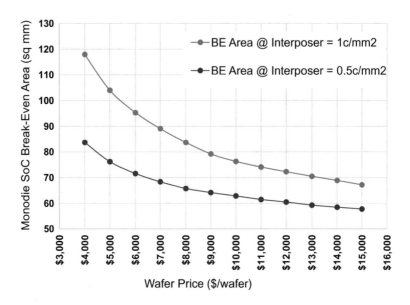

Fig. 2.19 Mono-Die SoC versus Split Die SiP Break-Even Locus versus wafer price—a plot of break-even locus of Mono-Die SoC area where cost of Mono-Die SoC is equal to the cost of Spilt Die SiP versus Si Wafer price, with the cost of the interposer as a parameter, and at a fixed Process Defect Density (D0 = 0.3 def/cm^2) and typical assumed values for other variables

technologies going down, at advanced CMOS technology nodes the favorable Split-Die product intersect takes place at smaller die sizes. This is illustrated in the Fig. 2.19—using simplified assumptions (D0 = 0.3 def/cm^2, homogenous equal die split, 40 um uBump pitch, etc.), i.e., as shown, as the price/wafer goes up, the break-even die area where cost of Split-Die SiP equals to the cost of a Mono-Die SoC, goes down. Thus the Split Die SiP Value Proposition scales with increasing Si wafer price associated with successive CMOS Technology nodes, and favors SiP implementation versus Mono-Die SoC at progressively smaller die sizes.

However, note that as the die sizes go down, some of the tradeoffs described below (Sect. 2.4) also change. Consider a case illustrated in the cartoon in Fig. 2.20 and summarized in the Table 2.1 below—for a 144 sq mm Mono-Die SoC split into 2 die SiP.

Assuming standard CMOS node n to n + 1 scaling factor (50%), and allowing for typical content growth (~ 10%), on scaling from generation to generation the area of a comparable Mono-Die goes down by ~ 40%. However, given all the other Area/Yield based Split-Die constraints (Aspect Ratio, uBump Pitch, Package keep out areas, etc.), the D2D edge on each of the two die also goes down (from 12 mm to ~ 8.5 mm in this example). Therefore, assuming equal partition that requires a fixed number of D2D pins (say 2500 wires), the number of columns of D2D I/O's must go up, and the line/space spec for the D2D interconnect must go down! Thus, with continued CMOS scaling, even if

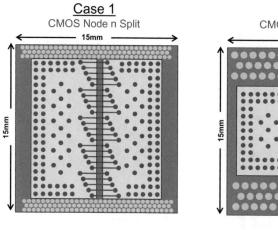

Case 1	Case2

CMOS Node n Split	CMOS Node n+1 Split

- Si Area Total ~ 144 sq mm
- Die Edge ~ 12 mm
- Package: 15 x 15 mm
- D2D Signals ~ 2500
- **D2D L/S ~ 2u**

- Si Area Total ~ 80 sq mm
- Die Edge ~ 8.5 mm
- Package: 15 x 15 mm
- D2D Signals ~ 2500
- **D2D L/S ~ 1u**

Fig. 2.20 Concept Split Die SiP with scaled CMOS Die—a cartoon illustration of a concept
2.5D Split Die SiP implemented in two successive CMOS Technology nodes, showing the impact
of scaling on die area and perimeter, and highlighting the consequences for the 2.5D technology
requirements such as the required die-to-die interconnect pitch

the number of D2D pins is constant (in fact all other things being equal the
number of D2D pins can be expected to go up due to truism described by
Rent's Rule), the D2D interconnect pitch must go down.

Therefore, for Split Die SiP implementation either the partition of the split has
to change at every node, resulting in successively more serial D2D interface,
or the SiP interconnect technology must keep pace with Si technology and
scale the density of wires per mm die edge every ~2 years as well.
*Consequently, for a SiP architecture that is expected to encompass successive
CMOS technology nodes, the Split Die versus Mono-Die tradeoff changes at
every node, and trends towards favoring smaller die areas requiring progressively tighter D2D interconnect technology.*

ii. ***Non-Scaled Split Die SiP Value Proposition***: Split-Die architecture offers an
opportunity for heterogeneous integration where different CMOS technologies
are used for different IP blocks. This is especially valuable with IP blocks—
such as for example PHY blocks—that do not lend themselves to standard
CMOS scaling (versus straight digital functions). One obvious, and potentially
easiest solution is to split the Mono-Die SoC and remove the PHY content
onto a separate co-located die—not for the area/yield advantages, but in order
to maintain the scaling benefits. With such a split, the PHY content is placed

Table 2.1 An example of scaling a Split Die SiP—a tabulation of die sizes for a specific example case that considers scaling a mono-die SoC and splitting it into a 2 die SiP, and illustrates the consequences for Split Die SiP SiP D2D interconnect. In the example the reference mono-die SoC in CMOS node "n" is a 12 × 12 mm die, and scaling it to n + 1 node with a 50% area shrink factor, and allowing for content growth of 10% in area, it ends up as a 8.9 × 8.9 mm SoC. A 2-die split for node "n" produces two 12 × 6 mm die, requiring 9 columns of D2D I/Os (for 2500 wire interface) and 2u line/space interconnect (for 40u uBump pitch). A same 2-die split for node "n + 1" produces two 8.7 × 4.5 mm die, requiring 12 columns of D2D I/Os and 1u line/space interconnect

Node	Mono-die SoC area mm2	Split die	Split die y mm	Split die X mm	Split die area/die mm2	D2D wires	Wires/mm die edge	uBump pitch um	D2D I/O columns	Target D2D Line/Space um
N	144	2	12	6	72	2500	209	40	9	2
N + 1	Area shrink factor 50%									
	72		8.5	4.24	36		294		12	1
	Content growth 10%									
	79.2		8.7	4.55	39.6		288		12	1

on a separate die, implemented in a given CMOS node, allowing the rest of the
SoC content to be scaled, with full area advantages. The value proposition is
then the reduction in the die size in the (expensive) node. Savings in die area
of few mm^2 on a very advanced node could translate to many $$ per KGD die;
savings that would then need to cover the cost of the PHY Die and SiP
package to make a positive Value Proposition.

An example of this value proposition is illustrated in the Table 2.2 and
Fig. 2.21, where Mono-Die SoC implementation in 3 successive CMOS nodes
is shown, with the PHY blocks highlighted in red. PHY blocks—used to
interface to external DDR memory, and other similar functions—basically do
not scale, and for a given set of interfaces tend to be approximately the same
area regardless of the SoC CMOS generation. This area is dominated by the
PHYs for the DDR interface—which is in fact dictated more by the
JEDEC DDR standards than by the actual CMOS technology node. In this
example, applying a constant area scaling factor (50%) and allowing constant
digital content growth (10%), PHY area on a sample SoC implemented in
node n (Area ~ 144 sq mm) is $\sim 7.0\%$, growing to 12% at n + 1 (SoC
Area ~ 84 sq mm) and $\sim 20\%$ in n + 2 (SoC Area ~ 50.5 sq mm).

The D2D interface for this kind of a split is estimated to require approximately
1000 signals between the digital and the PHY dice, but allowing for interface
overhead for PI and SI, control signals, etc. it is reasonable to assume
that ~ 1500 D2D wires are required. Maintaining a reasonable aspect ratio for

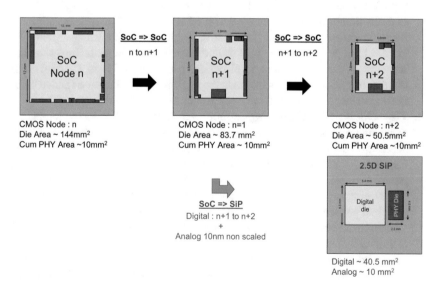

Fig. 2.21 Concept PHY Die SiP versus scaled Mono-Die SoC—a cartoon illustration showing
the effect of constant PHY area, highlighted in red, on scaled area of a concept Mono-Die SoC
implemented in 3 successive CMOS nodes. Also highlighted is a concept 2.5D PHY Die SiP
implementation where the PHYs are realized in a separate die, and integrated in a 2.5D SiP package

Table 2.2 An Example of a split PHY Die SiP—a tabulation of die sizes for a specific example case that considers scaling a mono-die SoC through 3 successive CMOS nodes, and illustrates the effect of non-scaled analog content (e.g., PHY). In the example the reference mono-die SoC in CMOS node "n" is a 12 × 12 mm die, with a 10 mm^2 analog content. The example is based on scaling the digital content by an area scaling factor of 50%, and allowing for 10% growth in content, while keeping the analog portion constant at 10 mm^2

CMOS node	Scaling factor	Content growth	Total die area (A) mm^2	Digital area mm^2	Analog Area mm^2	Analog content % of A (%)	Effective scaling factor
n	n/a	n/a	144.0	134.0	10.0	6.9	n/a
n + 1	50%	10%	83.7	73.7	10.0	11.9	41.8%
n + 2	50%	10%	50.5	40.5	10.0	19.8	39.6%

the PHY die, the long die edge for the D2D interface works out to be of the order of ~4–5 mm. Assuming a 40 u uBump pitch, this works out to be >12–15 columns of D2D I/Os, which (a) requires some Si area overhead that dilutes the value proposition, and (b) dictates the need for interconnect pitch of ~1.3–1.5 um. That is, implementation of this kind of a split, involving a very tiny PHY Die, in a single 2.5D type of a SiP package, definitely requires an advanced 2.5D technology with very tight interconnect pitch.

In fact, the complete tradeoff analysis for the example case used above is very complex, as all the overhead costs need to be comprehended, including the added D2D Interface area on both die (factoring in the under-bump utilization), the added cost of the package, the potential cost/savings in test and yield, etc. all as a function of the node (and cost) of the PHY Die and the SoC. In addition to the cost, an assessment of the relative (Split PHY Die versus Mono-Die) system-level performance impact should be performed, as well. However, since the long pole in the system level critical path (e.g., CPU to DRAM) is the number of cycles required to reach to the DDR memory, the few ns added by the D2D interface are expected to have a minor overall impact.

iii. **Tiny Die Access Value Proposition**: Advanced CMOS technologies, are expensive, and becoming more so—with costs per mm^2 of Si increasing in every successive node. This is driven by increasing processes complexity, driving CapEx up, and the mask count heading toward 100 masks. However, power-performance constraints favor implementation of high performance digital circuitry in advanced technology nodes. Hence, given that Si is the principal portion of the cost of an IC, business considerations dictate the use of minimum die size possible. But, for applications where the IC component ASP is limited by market conditions—such as mobile space (Chap. 5)—the die area has to be decreased in every successive CMOS technology node in order to compensate for rising cost per sq mm of Si. In this environment of decreasing die sizes, managing the access to the periphery of the die becomes a growing challenge and requires decreasing interconnect pitch associated with 2.5D technology.

The I/O count needed to interface SoC/SiP to the rest of the system can be expected to stay level and/or to continue to grow with successive CMOS generations. This is driven by considerations such as

- SoC Content and complexity is growing—to accommodate more cores and more core types, controllers, multiple RF bands, etc., driving modern mobile SoCs into $\sim 10B$ transistor range. Basic Rent's Rule dictates that this will drive an increase in the pin count.
- Power management is necessarily getting more complex—driving an increasing number of operating, power down, and power throttling modes, multiple power domains, etc., resulting in increasing number of power and control pins.
- System Memory density and bandwidth is growing—driven by larger diversity and count of applications, and by higher quality displays— resulting in the need for increasing number and width of memory channels, and increasing number of interface pins to system DRAM memory.
- The number of external devices—sensors and the like—that a SoC manages, is growing, thereby driving the need for increasing number of interface pins.
- etc.

The typical pin count for the modern SoC for a mobile phone product is of the order of $\sim 1000 +$ pins total, with ~ 500 signal I/Os and the rest as power/ground and utility pins. Figure 2.22 illustrates the trend in the total pin

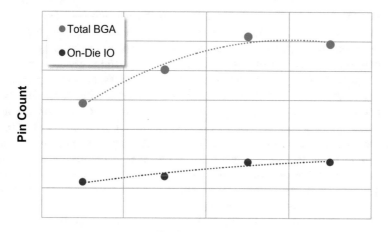

Technology Node

Fig. 2.22 Pin count trends for mobile SoC products—a plot of pin count of successive generations of a leading edge mobile SoC products implemented in last few CMOS Technology nodes, with a breakdown of on-die I/O count and on-package BGA count

count, as well as the on-die I/O pin count, for a mobile SoC, over the last few CMOS generations. As expected, both, the total and the I/O pin count is increasing—with power/ground requirements growing faster than the signal pin count.

The principal factors that constrain the external pin count are: (a) the package size and BGA pitch, and (b) die size and Cu Pillar pitch.

BGA pitch is ultimately constrained by the client PCB technology, and in the high volume commercial space (e.g., mobile) most of the OEMs seem to be quite reluctant to go below the current $\sim$0.3–0.4 mm. Hence, package size is constrained to somewhere around 15 mm × 15 mm de facto limit to accommodate the necessary BGAs (a fully populated BGA array on a 14 mm × 14 mm package accommodates $\sim$1000 pins with 04 mm BGA pitch). Note that use of external memory versus PoP memory is a big factor that can add/subtract few hundred pins to the total count of BGAs, and hence can affect the package size. Nevertheless, BGA count is typically managed by keeping the package sufficiently large to accommodate the necessary balls.

With die size of $\sim$10 mm on a side ($\sim$ 100 sq mm area) and with Cu Pillar technology with 80–100 um pitch, escaping the die is effectively removed as a constraint—including the usual PoP implementation, i.e., with 80–100 um Cu pillars die periphery of 10 mm × 10 mm die is adequate to house more pins than are needed. However, as the die sizes go down to $\sim$30–50 mm^2, and below, with successive CMOS nodes, die periphery is reduced and 80 um pitch becomes a limitation on pin count. Thus, when die sizes are shrunk in advanced CMOS technologies, use of fine pitch uBump technology (30–40 um) with a correspondingly fine interconnect pitch ($\sim$3 um and below) becomes a necessity. Alternatives, including (a) either fewer pins—at the price of power and/or performance, or (b) larger die—at a price of extra \$\$, are not palatable. In fact, a complete analyses of the tradeoffs between die area and periphery options is very complex. When the next level of details (such as the placement of PHYs and other features that have specific constraints on-die perimeter placement and pin counts, MFU factors that constrain die aspect ratios, types, and number of interfaces to external ICs, number of GPIOs, etc.) are all taken in account, as well as the packaging technologies capabilities (Bump pitch, Pad size, interconnect pitch), then it becomes apparent that a given packaging technology runs out of steam at a given die size for a given pin count, and capabilities associated with the advanced 2.5D Technology become a mainstream requirement.

Hence "high pin count + small die" products require a fine pitch die-to-package interconnect technology. Note that this constraint is true and as applicable to Mono-Die SoC as much as for a Split-Die SiP.

2.4   2.5D SiP Technology Tradeoffs

SiP cost–benefit analysis is extremely complex, and involves many, highly inter-dependent, tradeoffs. Significant engineering effort has to be invested to understand and quantify these tradeoffs, for each IC product, at every CMOS technology node, in order to identify the "sweet spot" where implementation of the 2.5D SiP package is better than a Mono-Die SoC. It is very much a matter of the proverbial "peeling of the onion" exercise, outputting product cost and performance implications for various interdependent technology, architecture, and the physical design variables. Some of the principal "knobs" for this kind of tradeoff analyses are discussed below.

2.4.1   Architecture Knobs

Partitioning of an architecture initially defined for a Mono-Die SoC into an architecture suitable for Split-Die SiP demands a complex tradeoff analyses. To the uninitiated and unsuspecting, it would appear that splitting a die that contains a dozen or two of given IP blocks, into two smaller die, would not be a big challenge. In fact, it turns out, that the implications of the various partitioning schemes impact SiP product cost, performance, power, latency, interfaces to standard ICs, and/or even software stacks. This analysis requires a series of what-if studies involving a multidisciplinary skill mix. Some of the key architecture knobs include:

i. *Die area*: in order to maximize the area-based value proposition, Split-Die SiP, using the same CMOS technology node for both die, requires that the two die are approximately of equal die sizes. The cost benefit of the split, as discussed above, erodes as the die sizes become asymmetric—since the larger of the two die would dominate the net yield of the pair. Note that if the partitioning leverages a heterogeneous option—where the two die are not implemented in the same CMOS technology node—then the optimum balance of the die sizes shifts and is dependent on the relative scaling factor and wafer pricing of the two nodes. This is illustrated in the Fig. 2.23 showing the relative savings in aggregated Split Die cost versus equivalent Mono-Die SoC as a function of relative area of the two die, with CMOS nodes as a parameter. Note that the selection of the CMOS technology node is constrained by power and performance requirements—not just the cost considerations. Thus, for example, the cost structure alone of older nodes, such as 28LP, may favor a heterogeneous SiP solution, even after compensation for the area scaling factor, but the modern nodes, such as 10 nm or 7 nm FinFETs, have com-pelling power and performance advantages.

Fig. 2.23 2.5D split Die KGD benefit versus area distribution—a plot of relative cost savings of Split Die SiP versus equivalent Mono-Die SoC, as a function of relative distribution of area between the two die, with CMOS nodes as a parameter. As shown, for a homogenous split peak savings corresponds to an even 50%–50% partition. Heterogenous split favors larger proportion of content in older and cheaper CMOS nodes

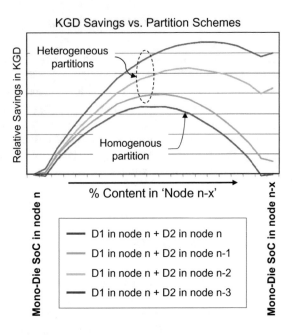

In any case, the partitioning of the mono-die SoC must result in approximately comparable die areas for each of the die, thereby constraining the possible distribution of the IP among the two die.

ii. **Die shape**: not only do the two die need to be roughly the same area, but they also need to be rectangular with an aspect ratio of about 2, i.e., die height should be $\sim 2x$ die width, as illustrated in Fig. 2.24 sketch. The driver for this is that the final product package should be close to a square—in order to be user-friendly with the PCB place and route requirements, and—in case of mobile applications—to maintain compatibility with PoP memory package.

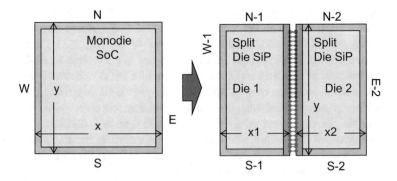

Fig. 2.24 Preferred Die shape for 2.5D SiP—a cartoon of a 2 die SiP package illustrating the favored Split Die configuration where the die are equal and have an aspect ratio of ~ 2 (rectangular die with die height $\sim 2x$ die width)

Rectangular die favor use of rectangular IP blocks (Sect. 2.4.2) in order to avoid possible on-die routability issues. Note that some of the IP blocks intrinsically have aspect ratio preferences and are better when roughly a square, or rectangular shape, due to access to shared resources, availability of suitable on-die I/O, etc.

Thus, the allocation of the IP blocks to a given die can be constrained not only by the overall area requirements, but also by the potential on-die packing density considerations.

iii. **External I/O Distribution**: furthermore, the external I/Os, i.e., those interfacing to the outside world (memories, sensors, drivers, other chips.), need to be distributed judiciously across the two die, in order to maintain package level routability. That is, if the I/Os that need to be routed to the package BGAs are clustered on one of the two Split-Die, then this results in a routing escape challenges that may force the package and/or the interposer to higher cost structure (more interconnect layers and/or finer pitch).

On the other hand, I/Os that interface to the PoP memory and need to be routed through PoP Vias, do need to be clustered on one of the two die, in order to interface to a single memory controller and to manage balanced routing to the memory.

Thus, the distribution of the IP across the two die is also constrained by the specific I/O distribution requirements, and package (and even PCB) routability considerations.

iv. **System Power-Performance**: the impact of the Split-Die partition on the overall system performance is clearly a key constraint. It is unlikely that the Area-Based Split-Die SiP implementation will perform better than a Mono-Die SoC—after all there is no way that the Split-Die 2.5D SiP will end up with shorter interconnect wires than what is possible in a SoC (as opposed to 3D stacking in Chap. 3). However, with a judicious Split-Die partition, the (negative) performance impact can be minimized and the cost benefit can be maximized. Defining and quantifying the impact of a given partition of a Mono-Die SoC into Split-Die SiP is a lot of work, and engineering analyses is required to select the right partitioning schemes. Hence, in order to manage the effort, in practice, some judgment has to be exercised to screen out unlikely candidates, based on generic estimates and/or architecture sanity checks.

The cartoon in Fig. 2.25 illustrates the range of choices. In addition to the basic blocks such as the CPU, GPU, Cache memory, and the Interface to external DRAM memory, modern Mono-Die SoCs typically include a couple of dozen specialized accelerators and IP blocks—such as for example, audio and video cores, display drivers, modems and radios, DSPs, sensor interfaces and associated processing units, utilities like clocks, on board thermal sensors, power management blocks, etc. Clearly, there is a number of different partitions which may comply with the geometric constraints described above, but each one of which will have a different impact on power and performance. Even after the major functions (GPU, CPU, Cache..) are allocated, and

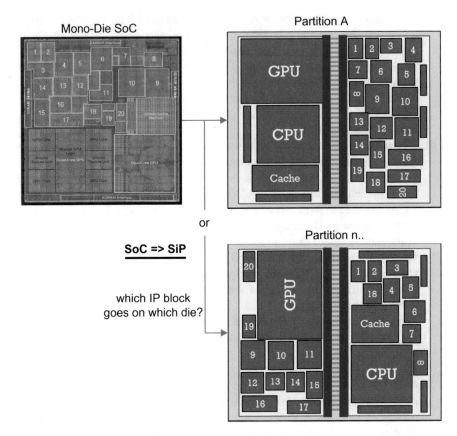

Fig. 2.25 Split Die SiP distribution of IP—a cartoon of a 2 die SiP package illustrating the range of possibilities for allocation of typical IP blocks found on a Mono-Die SoC (AnandTech die image) between the two die in a SiP

'stupid' alternatives are eliminated, the number of permutations allocating the ~ 20 other IP blocks to Die1 or Die2 in Split Die SiP is large, and estimating power/performance implications of each scheme is a lot of work. So either a structured PathFinding environment (Chap. 4), or a multidisciplinary team, or a mix of the two, is required to quantify the power-performance tradeoffs of the various possible partitions.

Some general considerations are identified below:

a. *Access to the system memory*: Clearly access to the external DRAM memory is critical to the performance of CPU, GPU, and several other IP types—and so to the overall system performance. Hence, the SDRAM Interface blocks typically need to be placed on the same die as the memory-performance sensitive blocks. Blocks placed on the other die would then incur some memory latency penalties associated with transfer of data across the D2D interface. Hence the die partitioning decision may be dominated by sensitivity to memory latency and bandwidth and which of the split die carries the DDR interface.

b. An alternative architecture giving each die direct access the system memory would maximize the partitioning degrees of freedom—allowing placing GPU and CPU and other memory-performance sensitive blocks on either die. However, this would require two sets of DDR interfaces (extra Si area, extra pins), and would precipitate coherency challenges. One way of offsetting the cost impact of dual DDR interfaces is to use a hybrid, non-unified memory architecture—e.g., combining hi performance LPDDR4 memory for die 1, with lower cost LPDDR3 memory for Die 2, and thereby reducing the overall cost of the DDR memory stack. However, enabling this would require upgrades to the Operating System to manage the priority and coherency of multiple memory stacks (see Chap. 4).

c. In principle, the degradation of the performance caused by DDR memory access limitations could be offset by deploying an extra level of cache, or a larger cache memory, and/or a cache memory for each of the split die. This too however, would then add extra Si area and would increase the AUC.

Thus, at least one parameter that dictates the SoC partition architecture, with major impact on system performance, is the sensitivity of each IP block to memory latency and bandwidth (i.e., latency sensitive blocks need to be placed close to the memory interface). Other dominant considerations could be access to shared resources (e.g., cache memory), or common resources (e.g., some sub-system bus), or separation of close-coupled blocks (e.g., CPU and Cache…). And so on.

v. **IP Availability**: Another constraint to be considered is the availability of the desired IP in a given CMOS technology. One of the key potential value propositions offered by a heterogeneous Split-Die SiP is enabling an option to NOT scale every IP block to the same, advanced, technology node that is required for some other IP block. Thus, the processing units (e.g., CPUs, GPUs) can leverage the power-performance advantages offered by the bleeding edge CMOS node, and competitive pressures force implementing these blocks in the latest node possible. However, other blocks (e.g., Audio or Multimedia…), may not necessarily have the power and/or performance benefits from an advanced node to justify the costs, and some IP blocks (e.g., analog blocks) may actually have a disadvantage when scaled to the most advanced node. In addition, some IP blocks may not exist in older nodes, and investing the design effort to create a given function in an old node does not have the reuse and portability opportunities associated with designing it in a new node. Thus, IP blocks—such as for example the LPDDR4 interface, or the latest LTE Modem, etc., do not exist in older CMOS technology nodes— e.g., 28LP—and therefore limit the choice of CMOS nodes for each die.

> *Therefore, the partitioning choices and allocation of IP blocks to either of the*
> *two Heterogeneous Split-Die may be constrained by the availability of the*
> *right IP and/or the design effort tradeoffs associated with designing new IP.*

vi. **D2D Interface Performance**: There is clearly a tradeoff associated with the design of the D2D Interface itself. A serial interface requires few pins, but typically costs power and latency, and requires incremental design effort (versus standard SoC) at the architecture, or even software, level. A parallel interface is power and delay efficient, and is compatible with the architecture used on a Mono-Die SoC, but requires many D2D pins, thereby driving up the cost of the 2.5D technology solution. Clearly there is a tradeoff, and a "sweet spot" is dependent on product, partition, technology, and so on. For a given partition, target system-level performance requirements define the spec for the inter-die communication bandwidth for the Split-Die D2D interface. D2D I/O and channel (wiring in the interposer) define the maximum frequency of the D2D pins. Conceptually, these two constraints then define the number of the D2D signal pins needed to provide the necessary performance (an additional set of pins needs to be allocated for the various utilities needed to synch up the two die, power, grounds, etc.) and hence constrains the 2.5D packaging technology options and many other cost tradeoff knobs.

Note that the architecture of the actual D2D interface can be based on some type of a bus architecture. A bus protocol then needs to be defined to manage the handshake between the die, and to synch up with the rest of the system busses and communication structure, etc. That is, D2D Interface architecture can be quite complex, in itself, and again, the sweet spot is product and technology specific.

> *Thus, the number of the D2D pins is a fundamental tradeoff variable, and is*
> *one of the key parameters driving the overall cost and performance of a Split*
> *Die SiP.*

That is, even at the fairly basic architecture level, the partition of a Mono-Die SoC into a Split-Die SiP is bound by many constraints. It is not trivial at all.

2.4.2   Physical Design Knobs

Physical design of the Split-Die also involves different constraints than applicable for the traditional Mono-Die SoC. These can be segregated into the challenges associated with general physical design of the two die and the design of the D2D interface itself. Specifically:

i. **Design Methodology**: a key new feature introduced by the Split-Die architecture is the D2D interface itself—which in turn imposes constrains for the design methodology. Given the various possible challenges and penalties, the most practical solution for the design of the D2D interface is to leverage some suitable standard bus protocol to manage the data transfer across the interface and the handshake between the die.

Using a standard bus interface allows the design of each of the two die to be implemented separately—the "two-separate-designs" approach—with the Bus Protocol dictating the handshake and timing constraint on each of the die. The "two-separate-designs" approach is the easiest, but eliminates some flexibility in partitioning—for example, each die must include all the blocks necessary for its timing closure (also for verification at the die test level). In addition, the generic nature of any bus protocol necessarily dictates use of excess margins—for example the timing specifications for the interface must be sufficiently conservative to be valid across all possible corners (process, temperature, voltage). Thus this approach is potentially sub-optimal from the partitioning and performance point of view. In addition, the design of the interposer and the D2D interconnect has to be done separately too, with quasi-manual interface to the active Si die design—thereby also introducing some potential inefficiencies.

On the other hand, current generation of design tools would be challenged by any approach that involves two die + the D2D wires in the interposer, within a single design. Managing, for example, timing across different process corners of the two separate die in a single design, would be difficult. Even if it was enabled in the EDA tools, the standard "corners" methodology would explode with the number of corners that would need to be characterized. Conversely, dealing with the logistics required to ensure that both die come from the same process corner is also quite unpalatable. In addition, importing a separate interposer and/or package technology file to deal with the delay of the relatively long D2D interposer wires, into a single design, would be complicated.

Thus, use of "two-separate-designs" approach, with some kind of an D2D Bus Spec and D2D I/O Spec, is really the only practical solution. Standard design practices, tools and flows, constrained by specifications dictated by the D2D bus protocol, in addition to the usual interface standards (e.g., DDR, PCI, I2C, *etc.), I/O constraints (Drives, Levels, ESD.) and loads (R-L-C of the package and PCB wires..), can be used—along with the inefficiencies associated with this methodology.*

ii. ***Physical Design (PD) for Split-Die SiP***: Mono-Die SoC design typically favors a floorplan that results in a die that is approximately square, with the hard macros distributed around the periphery of the die (inside the donut defined by the I/Os, PHYs, and other blocks that require tightly controlled access to the periphery), and the top level inter-macro interconnect routed through the center of the die. This typically results in reasonably good uniformity and utilization efficiency. Split-Die implementation on the other hand favors two, roughly equal sized, rectangular die with aspect ratio ~ 2.0, in order to end up with the combined die form factor that is close to square and fits in a square package (Sect. 2.4.1).

This precipitates some new potential PD challenges. Clearly, in order to manage the KGD costs, the combined die area must be kept close to equal to the corresponding Mono-Die area. However, implementing a routed design in a floorplan with high Aspect Ratios results in higher congestion for the vertical routes. This is especially so if the aspect ratio of the hard blocks is not adjusted to sufficiently rectangular shape, so that the routing along the y-axis is bottlenecked between the hard macros.

In addition, access to the perimeter of the die is complicated with higher aspect ratio floorplans. Assuming that the Split-Die partition has been defined such that the number of external I/Os is approximately balanced and equal for each of the die, they clearly have to be distributed across the 3 sides of each of the die, since the fourth side has to be dedicated to the D2D interface. Some interfaces—e.g., DDR memory—need to be laid out with reasonably balanced wire lengths, so that they typically are not wrapped around the die corners. But, if the North/South (see Fig. 2.24 for description) die edges are not long enough to accommodate the DDR interface, they may be forced to use the fourth side (West on Die 1, East on Die 2). This may then dictate the need for a new layout of the interface IP to accommodate changed poly orientation, and pushes other I/Os to the North/South sides. On the other hand, if N/S side is long enough then the rest of the I/Os are forced to use the W/E sides, possibly compounding the routing challenges at the package level.

Altogether, the higher aspect ratio floorplan required by the Split-Die design, ends up impacting routing efficiency and efficiency of utilization of the die perimeter. This has to be comprehended in die size estimates and the general 2.5D tradeoff analyses

iii. **Physical Design of the D2D Interface**: This is clearly a challenge specific to Split Die SiP. The total number of D2D pins required is dictated by the necessary bandwidth needed by the D2D interface, and the maximum acceptable frequency of the D2D pins. The uBump pitch and the Y dimension of the Split-Die, in turn, dictates the number of D2D pins that can be accommodated in a single column along the die edge. Combination of these requirements therefore dictates the number of columns of the D2D I/Os required to implement the entire interface. Thus, for example, if the bump pitch is limited to ~ 40 um, then only ~ 25 bumps can be realized per mm die edge. In order to get to the desired 100–200 wires per mm, the D2D I/Os have to be placed in multiple columns (4–8 columns in this example), and the interposer interconnect pitch has to be sized to allow routing between the bumps to access the inboard columns ($\sim 6 - \sim 3$ um pitch for this case). This is illustrated in the layout cartoon in Fig. 2.26.

iv. **Area of the D2D Interface**: Note that the uBumps required to accommodate the D2D interface is an incremental feature relative to Mono-Die SoC, and therefore incremental Si area on each die is required to accommodate the total number of D2D uBumps. The D2D I/Os are placed in the area under the bumps. The D2D I/Os are quite small and could take an area that is lower than the area

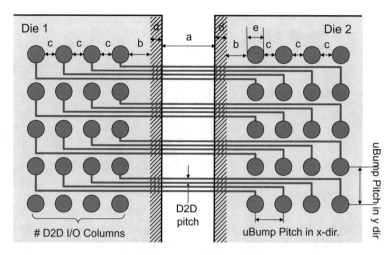

Fig. 2.26 Physical design of 2.5D Die to Die interface—a cartoon of a 2 die SiP package illustrating a zoom-in on portions of the two die, and highlighting the critical features (uBumps and D2D Interconnect), and dimensions (uBump pitch, D2D Interconnect pitch, Keep Out Zones)

of the uBumps (40 × 40 um). However, it is difficult to effectively use any leftover white space under the peripheral ubumps, since it is by definition located along the extreme die edge—where there is low demand for either logic placement space or interconnect routing space.

Thus, the incremental Si area that has to be added to each die for Split-Die implementation is typically dictated purely by the uBump count and placement rules

$$\Delta A = (\# \, of \, pins \, per \, column \, * \, uBump \, pitch \, in \, y \, axis)$$
$$* \, (\# \, of \, columns \, * \, uBump \, pitch \, in \, x \, axis).$$

v. *Electrical Design of the D2D interface*: The length of the D2D interconnect wires (from uBump on Die 1 to uBump on Die 2 through the interposer RDL) is clearly a function of the keep out rules and die to die spacing, as well as the number of columns of the D2D I/Os. With typical spacing and uBump pitches, the average D2D wire length ends up at approximately ∼1 mm long. Whereas 1 mm, or longer, wires are common in package design, they are rare in Si design. Furthermore, the long package wires are implemented in relatively thick copper wires on a ∼20u pitch, and are driven by large I/Os facing the outside world. With 1 mm long wires that are a few um thick and on a 3 um to 6 um pitch driven by low drive I/Os, used for the 2.5D technology, Signal Integrity (SI) and Power Integrity (PI) become issues that must be considered. Signal integrity is driven by wire inductance and signal edge rates (L * di/dt) and is exacerbated by any erosion of power voltage. Normally SI and PI are mitigated by minimizing Vdd droop, and by providing grounded shielding—both of

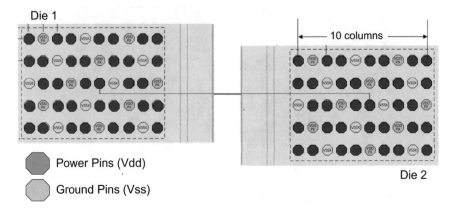

Fig. 2.27 Electrical design of 2.5D Die to Die interface—a cartoon of a 2 die SiP package illustrating a zoom-in on portions of the two die and highlighting the Power (red) and Ground (blue) bumps distributed across the D2D signal (black) array

which call for very good local power and ground planes for the D2D I/O interface.

Aside from the geometrical considerations that control wire inductance (L) and capacitance, Signal Integrity is determined by the I/O driver strength (di/dt—dictated by the frequency requirement), and driver effective load (dominated by the ESD protection). Hence ESD protection level is a trade off variable. Since the D2D pins are not wired to face the outside world, they are not expected to experience true ESD discharges caused by handling, and hence must not necessarily carry the standard Human Body Model (HBM) level of protection. On the other hand, the D2D I/Os can be subject to discharge events during the assembly process, or even induced by some of the plasma process steps. In the absence of any definitive data that defines the frequency or the level of discharge that the die may be exposed to during the 2.5D technology processing, the "safe" engineering judgment is that the D2D I/Os need to have immunity for up to few 100s V of Charged Device Model (CDM) events.

With this ESD load, and assuming an I/O operating frequency of the order of 2 —4 GHz, and with 1 mm D2D wire lengths, Signal Integrity requirements (overshoot, eye opening, jitter...) typically dictate a need to provide local power pins. In addition, ground wire shielding of the signal wires may be necessary to mitigate coupling, so that incremental D2D ground wires may be required (typically of the order of 1 ground wire for every ∼2–4 signal wires). Thus, incremental Power and Ground bumps may need to be distributed across the D2D interface array. The net D2D bump array layout then ends up looking as illustrated in Fig. 2.27—showing the power (red) and ground (blue) bumps sprinkled throughout the array

Note that the exact Signal Integrity (SI) and Power Integrity (PI) behavior is a function of many detailed layout decisions, specific use case and data traffic, the technology and material attributes, PDN and Power specs, etc. However, it is clear that SI and PI considerations dictate a need to add a significant number (∼ 20%) of incremental pins in the D2D array—as illustrated in the figure. Therefore, the Physical Design and Electrical Design of the D2D interface is an iterative interactive process where physical design affects electrical performance, and vice versa.

vi. **Design Overheads**: Split-Die architecture may require incremental overheads in Si implementation other than just the D2D interface described above. Since each of the Split-Die must be testable prior to assembly, there are the usual chip "utilities"—such as clocks and clock controls, DFT management circuitry, power management circuitry, potentially circuitry necessary to ensure security and privacy, fuse and fuse control circuitry, temperature sensors, etc.—that cannot be shared between the two die. These features therefore need to be replicated on each of the die. The exact magnitude of this type of overhead is hard to estimate—and clearly depends on the content of the SoC to be partitioned, and the specific partitioning architecture, product specs, etc.
Note however, that with continued Moore's Law scaling, these "design overheads" may become more significant, as they may involve portions of the circuitry that do not necessarily scale well into very advanced nodes (analog, I/O, etc.), and are the basis of the 2.5D technology benefits that are incremental to the area-based value proposition (Sect. 2.3.3).

That is, there is a perfectly good reason why the industry pursued higher levels of Mono-Die SoC integration for the past few decades—for a given technology node this does result in implementation that makes the most efficient use of Si area. Split-Die will always, necessarily, lead to some Si area overhead—both of the Design Overhead type and the D2D Interface type. Clearly these overheads must be considered when exploring the Mono-Die SoC versus Split-Die SiP tradeoffs.

2.4.3   Si Technology Knobs

The cost of the Si die is a major portion (typically ∼ 2/3) of the overall SoC Component AUC, and as such should be the biggest knob to drive the Split-Die tradeoffs. Certainly, as described above, the Benefit side of the equation is entirely derived from the Si domain, i.e., the relative benefit of yielded Split-Die Si versus yielded Mono-Die Si is the primary value proposition. Principal knobs that define the magnitude of that Known Good Die (KGD) benefit are:

i. **Wafer Price**: the fundamental price of a fully processed wafer depends on the technology node, number and type of masking layers, technology maturity and buyers' negotiating power. Note that with a homogenous Split-Die where the Mono-Die SoC and both of the Split-Die are all the same technology node and use the same number of masking layers, the relative impact of the wafer price on KGD benefit cancels out, since it is common to all sides of the equation. *However, wafer price is a significant constant multiplier that impacts the final* **absolute** *AUC, since the $ value of the Split-Die KGD benefit scales with the*

Fig. 2.28 Cost/Benefit of
Split-Die SiP relative to
Mono-Die SoC versus defect
density—a plot of cost (-ve)/
benefit (+ve) of 2.5D Split
Die SiP relative to Mono-Die
SoC versus Process Defect
Density, with the Wafer Price
as a parameter, and using
typical assumed values for all
other variables (die size, cost
of package, overhead area,
etc.)

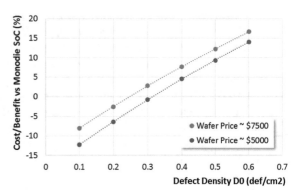

wafer price, and balances the cost of the SiP implementation, thus driving the net AUC benefit.

ii. **Defect Density**: defect density drives yield, and hence the relative benefit of Split-Die versus Mono-Die. This is illustrated in the chart in Fig. 2.28, showing a locus of break-even points between Mono-Die SoC and Split-Die SiP versus D0. Whereas the values are dependent on a given set of assumptions, the trend is generic, i.e., for a given die size, with low D0, Mono-Die can be more cost effective than a Split-Die. Note that there are many variables that affect the total real die yield. For example, actual yield is a combination of defect driven functional fails, systematic design-process interactions, and parametric driven levels/performance fails. With modern technologies the systematic and the parametric fails can very well be dominant—and the Area-based Split-Die benefit can be entirely washed out. Also, there is typically a radial dependence of D0 across the wafer, with the very center and outside edges having higher D0 than the donut region in between. The impact of this donut region is different for large die versus small die, and can amplify the benefits of Split-Die—sometimes significantly so. And so forth.
In general, however, the greater D0, i.e., the less mature that a given CMOS technology is, the greater is the benefit of Split-Die.

iii. **Yield Learning Rate**: Normally D0 goes down with process learning, and (hopefully) by the time that a high volume product is in manufacturing, Poisson D0 is down to less than ~ 0.3 def/cm2. Typically, yield learning is driven by the number of wafers processed, i.e., volume ramps drive yield learning. Split-Die architecture could either slow down the yield learning rate, due to a more complex implementation, or maybe speed it up, due to easier fault isolation.
Clearly yield learning rate is a key variable for estimating the total, averaged, tradeoff between Split-Die costs and benefits, aggregated over the entire product life cycle. Note that repeated studies (e.g., Leachman 2002) have shown that Learning Rate is an attribute of a foundry culture, and as such encompasses many more aspects of an organization than just the fab cycle time.

iv. ***Technology Node***: one of the potential benefits of Split-Die architecture is to allow heterogeneous integration. The performance and schedule tradeoffs associated with heterogeneous integration are discussed elsewhere (Sect. 2.3 and 3.3), but there is also a cost implication. If one of the two Split-Die do not require the most advanced Si node, or require fewer layers of interconnect, or otherwise dictate a different cost constraint, then the cost structure for that die could be lower—even after adjusting for the area scaling factors—resulting in lower overall KGD cost.

Heterogeneous integration then gives the Split-Die SiP a different option to optimize its cost structure—an option which Mono-Die SoC does not have.

v. ***Die Size and Shape***: as discussed (Sect. 2.3.2), the area-based value proposition of the Split-Die implementation is maximized when a Mono-Die is symmetrically partitioned into two approximately equal sized, rectangular die. Both the die size and die shape of the resulting die can impact KGD cost via the packing efficiency of a mask reticle, and/or optimizing the wafer area utilization and stepper throughputs (Mask Field Utilization factor: MFU).

Thus die aspect ratios can be a quite significant in the overall cost structure—especially with newer technologies where lithography steps are a dominant portion of the overall wafer processing cost structure. Note that this could have either positive or negative effect on cost of Split Die SiP relative to Mono-Die SoC.

2.4.4   Packaging Technology Knobs

Whereas Si technology is the primary driver of the benefit side of the SiP versus SoC cost–benefit equation, the packaging and assembly technology is the principal driver of the cost side. 2.5D Technology required for assembly of a Split-Die SiP does dictate requirements for various advanced packaging technology features—all of which ultimately add to the cost of the overall package.

Note that packaging technology is very rich in choices—more so than Si technology—offering many options in terms of core technology types, materials, processes, etc. All these options have different tradeoff points. The intent here is to focus on the tradeoffs pertinent to the 2.5D technologies—more specifically for the candidate technologies identified above (Sect. 2.2). The principal knobs include:

i. ***Interposer***: the specification of the target interconnect density—including line/space and via/pad dimensions, is the primary driver for the selection of the basic 2.5D technology. Conventional organic substrate technology is currently at about ~ 10 u line/space with ~ 75 um via/pad pitch. If higher interconnect density is required—as would be the case for a typical SiP architecture—then alternative technology candidates have to be identified, at least for the D2D interconnect layers that require tight pitch. The range of choices is illustrated in the chart in Fig. 2.29, showing routing pitch versus cost relationship for the

key technology candidates described in Sect. 2.2. Note that this chart is very
much simplified, and that many more than the 2 variables identified (routing
pitch versus cost/unit area) affect the tradeoff—but the trend, and the relative
cost points are generally appropriate. Note also that the cost estimates change
over time with process learning and/or future technology developments—for
example moving the Fan Out technologies to panel-based format could radically effect the relative positioning of that technology on the chart.

Interposer technologies that rely on use of organic dielectrics with
Semi-Additive Process (SAP) interconnect (Fan Out, POI, and LCIg in the
chart) probably bottom out somewhere between 2–5 um line/space specification, but are *currently* expected to be capable of hitting ~ 1 c/mm^2–1.5 c/mm^2
cost point in high volume manufacturing. Si interposer-based technologies
using a few (1–3) dual-damascene metal layers (TSI, LCIs) are easily capable
of supporting much higher interconnect densities (0.5–1.0 um)—but are
expected to be more expensive at $\sim >$1.5–2.0 c/mm^2 for the interposer, even in
high volume. Hybrid Technologies should be able to have cost comparable to
FO WLP, but with interconnect density comparable to LCIs/TSI. Glass-based
Interposer technologies could intrinsically be capable of high interconnect
density but the Through-Glass-Via (TGV) density, handling, film adhesion,
and sourcing issues are still to be resolved, so that the cost and risk is still to be

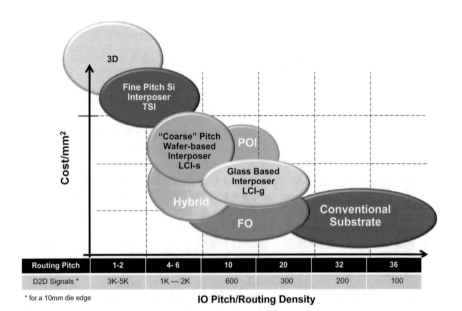

Fig. 2.29 Cost of Baseline package technologies versus Routing Density a chart illustrating
typical cost of various baseline package technology options versus their corresponding routing
pitch. Note that this chart is very much simplified, and that package cost structure varies with many
variables other than just the routing pitch and that the effective Routing Pitch depends on variables
other than just line width/space

fully understood. And finally, there is the fine pitch Si Interposer (TSI) with multi-layer (3 or more) dual-damascene metal layers for really high density interconnect, but at a pretty hefty price (~ 3 c/mm^2 or more).

Thus, the interconnect density driven by the D2D interface specifications drive the selection of the interposer technology, and therefore impact the overall packaging cost structure. Straight tradeoff between cost and density.

ii. **Die Connect**: the other key technology attribute that constrains the effective D2D interconnect density is the die bump pitch. Bump and I/O pitch has to be matched with the interconnect density in order to achieve the target D2D wire count (Sect. 2.3).

A bump pitch (and the interposer interconnect pitch) that enable implementing the full D2D interface with a single column of D2D I/O's is obviously the simplest solution. This would require least amount of overhead Si area, and would hence be the lowest cost solution, and it could be implemented with shortest wires, and would therefore require least effort in managing Signal Integrity. However, 5–10 um bump pitch, necessary to enable the 1000–2000 D2D wires on a 10 mm die edge, is currently not achievable in volume manufacturing. The minimum uBump pitch currently in production is 30–40 um, with 80–120 um pitch with the Cu Pillar technology and ~ 150 um pitch with the solder ball technology, being more of a mainstream solution.

The cost of packaging and assembly increases, in some cases significantly so, with reduced bump pitch. This is illustrated in the concept chart in Fig. 2.30— for a given set of assumed input variables ($/wafer, D0, Die Size, Package cost...)—just for purposes of illustrating the dependencies and trends. Low uBump pitch requires high density interconnect and hence an expensive

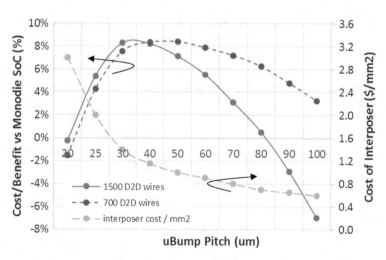

Fig. 2.30 Cost/Benefit of split Die SiP relative to Mono-Die SoC versus uBump pitch—a cartoon plot of cost ($-$ve)/benefit (+ve) of 2.5D Split-Die SiP relative to Mono-Die SoC versus uBump pitch, illustrating the concept that there is an optimum uBump pitch for a given set of constraints (using typical values for all other variables inc. Die Area, Cost of Package, D0, etc.)

interposer technology. Low uBump pitch also drives very tight requirements for planarity and warpage, and may force the use of Thermal Compression Bonding (TCB); which is an intrinsically more expensive process than the mass reflow bonding. In addition, use of tight pitch uBumps and TCB have implications on the specs for the pick and place process, and choice of the underfill materials and process, all of which in turn have implications for package size and cost. Finally, probing on small uBumps is difficult—current consensus is that it is not practical in HVM with 40–60 um uBump pitch (Sect. 2.4.5), thereby driving up the cost, and potentially eroding the quality, of test. On the other hand, loose uBump pitch requires larger incremental overhead Si area on each of the die (Sect. 2.4.2). Thus there is a 'sweet spot' for a given set of conditions.

Consequently, selecting the uBump pitch is in itself a tradeoff, with significant implications on the overall cost, and that can be optimized only in the context of the entire solution.

iii. **D2D Gap**: the separation between the two die on an interposer—typically of the order of several 10s of um's—is not a major contributor to the overall package form factor, but is important for the electrical performance of the D2D interface, and is controlled by various package technology choices.

Thus, as illustrated in the cartoon in Fig. 2.26 that illustrates a zoom-in on a section of the edge of the two die in a Split-Die SiP, a D2D interface that requires multiple columns of D2D I/Os, results in an average wire length that is controlled by die-to-die separation (a), die design edge to uBump keep out rule (b), uBump to uBump pitch rules (c), as well as the die design edge-to die physical edge gap (d). Whereas, from the electrical performance point of view, it is obviously desirable to keep the D2D wire length to a minimum, from manufacturability and yield point of view some spatial margins are required. Dimension "a" is an important factor for the underfill process. Dimension "b" allows for seal rings and other keep out rules for the die edge. Dimension "c"—as discussed above is a core uBump technology choice. And dimension "d" is actually an uncontrolled variable dependent mostly on the wafer sawing process. Thus, a variable that is normally considered a second-order factor—the width and consistency of the saw cut—is a first-order factor that impacts the D2D wire length, and some of the SI considerations.

Consequently, there is a tradeoff that balances cost and yield, favored by increased spacing and keep out rules, versus electrical performance, favored by minimized gaps.

iv. **Package Form Factor**: the target x-y and z dimensions of the final IC package, is clearly a fundamental constraint, with significant implications for 2.5D technologies. For mobile market, in practice, maximum package size is of the order of up to $\sim 15 \times 15$ mm package; albeit smaller is always better (but not necessarily monetizably so). This is driven by compatibility with standard PoP Memory package sizes (ubiquitously used in mobile phones), as well as the real estate and routability considerations for the phone PCB. In addition, the z-dimension is currently typically targeted at ~ 1.0 mm (including the PoP Memory), in order to squeeze all the components of advanced smartphone (frame, display, thermal solution, skins, battery volume) into an end product device that is fashionably thin (currently ~7 mm).

Since these constraints obviously apply to a Mono-Die SoC as well as the Split-Die SiP solution, and assuming that (judicious) partitioning into Split-Die does not add a lot of area overhead, form factor should not be a primary 2.5D Technology "knob". However, in fact, package target form factor, and especially the z-dimension constraint, is a limiting spec, forcing the use of very thin interposers and oftentimes exacerbating all other challenges. Thus, for example, with Si interposers, the easiest solution is to stack the interposer on a conventional substrate, in order to leverage the best attributes of either technology—as is done in TSI. However the thickness constraints prevent the use of this approach for mobile applications and force the use of Si Interposer attached directly to customer PCB (LCIs), with all the associated incremental challenges. Similarly, managing warpage and stress issues could be facilitated using thicker die (>> 100 um) for Fan Out technology candidates, and/or thicker core layers for POI technology—but may violate the z-dimension limits.

On the other end of the spectrum, the relatively large x-y footprint exacerbates managing the warpage, stress, Chip Package Interactions (CPI–see Chap. 4) and Board Level Reliability (BLR). Warpage and eCPI are challenges especially for the Fan Out technology which is intrinsically thin, while BLR is the key concern for Si Interposer and LCIg technology, which intrinsically do not match the CTE of the PCB.

Thus, the package form factor is a key constraint—especially for the mobile and IoT Wearable market—that must be factored in the tradeoffs. For other applications looking at 2.5D Technology solutions—such as Servers, PC, Games, etc.—package form factor is less of a constraint, and adding to package thickness and/or footprint size is typically weighed purely on the cost bases.

vi. **PoP Via**: Incorporation of connections for the PoP Memory—more or less a standard requirement for the mobile market—also poses a significant incremental challenge for 2.5D technology implementation. The challenge is harmonizing the x-y package size constraints with PoP Vias required for electrical connection to PoP Memory. Thus, incremental process to implement PoP vias with tall, plated Cu posts may be required if the x-y constraints do not allow use of conventional laser defined Through-Mold-Via (TMV) process. Similarly, for Fan Out technology, realizing PoP vias may require deployment of skinny "pcBar" slivers, posing incremental demands on the Pick and Place process and obviously increasing the complexity of the overall package. Furthermore, a PoP-side redistribution layer, enabled by either plating an incremental RDL layers on the topside of the package or by stacking an incremental PoP-side substrate layer, may be required.

That is, compatibility with PoP memory and corresponding implementation of a PoP Via process that is compatible with the rest of the 2.5D technology constraints may add significant complexity—and cost—to the final SiP package.

2.4.5   Test Knobs

2.5D technology with the Split-Die implementation also poses a number of challenges for test—both at the wafer and package level. Good "observability"—including stuck-at and timing fault coverage, Iddq coverage, and I/O parametric coverage, is essential for cost effective manufacturing. Screening out the manufacturing defects at wafer sort test is especially critical for managing the final AUC. That is, probe test at the die level is required to identify the KGD prior to packaging. Nominally there is a tradeoff between cost of the wafer sort test and cost of packaging—but for the 2.5D SiP components, with the relatively expensive packages and the possibility of assembling a good die with a bad die, sort test is "a must have." Therefore each of the die must be sufficiently operational on its own—at wafer and at package level—to enable good test and engineering de-bug.

Use of small chip attach bumps—especially uBumps with small diameter—poses probing challenges, so that there is a tradeoff between the uBump pitch versus the ability to probe each of the die. With uBump pitch below about ∼60 um, wafer probing for sort test in HVM is currently impractical. It is technically possible to probe on 40–60 um uBump pitch; but mostly for engineering purposes. The cost of the probe cards is prohibitive for HVM, and probing tends to be unreliable. Most manufacturing probe test, and the associated technology for building probe cards at a reasonable cost, is targeted for around ∼100 u bump pitch. Hence, with uBump pitches required for cost effective D2D interface, some kind of a SCAN based DFT solution must be deployed—where the observability is obtained even with fewer I/Os. DFT however, also brings overheads, and comes at a cost of some incremental Si area, and possibly somewhat compromised observability.

There is an opportunity to split the difference, and optimize the different overheads, by using multiple bump pitches. This can be done in manufacturing—for example current Mono-Die SoC products routinely use different Cu pillar pitch for I/Os and for Power/Ground (P/G) pins—as long as the diameter of the Cu bumps is kept constant in order to keep the height of the plated pillars (or solder balls) fairly uniform. If all the I/Os other than the dedicated D2D I/Os, are probable—and hence pitch compatible with probing technology—then a reasonably good test observability can be achieved for the chip logic, as well as for the I/Os that face the outside world. This then leaves the challenge of test coverage for the D2D interface itself. It is possible to define a fairly efficient DFT solution to assess the integrity of the D2D interface via a separate SCAN chain, and/or by probing on every second or third uBump (thereby effectively achieving 80 um or 120 um pitch probing with 40 um pitch uBumps). This calls for multiple Bump pitches; a fairly large one—of the order of 100u—for the external I/Os that face the outside world, a tight one—of the order of ∼40 um—for the D2D interface, and a loose one—of the order of ∼250 um—for power and ground pins, as illustrated in the cartoon bump image in Fig. 2.31. Note that with this combined approach—where Cu pillars are placed on different pitch for external I/Os versus D2D I/Os versus Power/Ground pins, it is important to ensure that there is access to good Power and Ground pins during

Fig. 2.31 Concept Bump map for a 2.5D SiP package —a cartoon illustration of a 2.5D SiP Package showing a concept die bump map, with a different pitch for I/Os, for D2D Interface and for Power/Ground pins

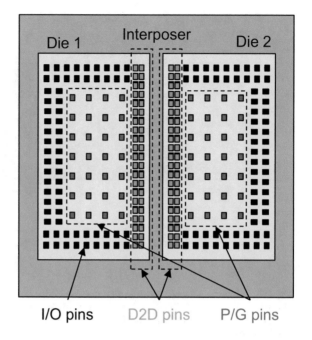

probing itself—in order to avoid undue Vdd droop during sort test of the D2D interface.

    Hence, there is a tradeoff between quality and cost of test—especially probe test —and die area and cost of the split die SiP.

2.5 2.5D SiP Technology Solutions

There are no universal, industry standard, set of specifications and requirements for the 2.5D technology. Thus, typically specific candidate technologies are selected for a given application to maximize a given value proposition. This section presents the typical selections for given value propositions, and discusses the general rationale behind the choice:

2.5.1 2.5D Technology Solutions for Integrated SiP

For the purposes of completeness an outline of a typical specification for the 2.5D technology targeting Memory Integration SiP Value Proposition is illustrated in Fig. 2.32.

HBM Integration 2.5D SiP Technology Spec	
Item	order of magnitude
Number of Interconnect Layers	3+ interposer / 8+ substrate
Interposer Line/Space	> 2u
Interposer Size	<~ 25mm x 35mm
Substrate Line/Space	~10u + (standard)
Substrate Size	~ 55mm x 55mm
Logic Die Size (max)	~ 25mm x 25mm x 0.7mm
HBM Stack Size	< 12mm x 7mm x 0.7mm
Package Size	> 55mm x 55mm
PoP	NO
Lid & HeatSink	YES

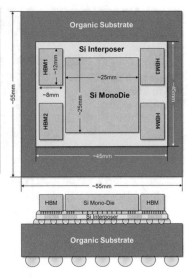

Fig. 2.32 Generic concept specification for the integrated memory 2.5D SiP—a cartoon illustration of a concept Through-Si-Interposer (TSI) SiP construction, showing a cross-sectional and a top-down view of an Integrated GPU + HBM package, and including a tabulation of typical target specifications

This is not the "worst case specification" for this application space, and, as indicated, 2.5D Technology for this class of SiP applications is already in commercial domain, and is hence not the focus of this book.

The 2.5D technology solution for this space is the TSI candidate. This approach has in fact been selected by most commercial entities competing in this space. Hybrid technology solutions that use "TSV-less Interposer" constructions with an organic substrate (SLIT) is also an alternative solution and in the long run, may dominate this space. Alternative potential candidates could be POI Technology—which is compatible with the large footprint requirements, but may be challenged by the very dense routing pitch requirements.

In the long run, it is believed that as commoditization of SiP technologies pushes ASPs down, TSI technology will migrate toward higher value propositions, where increasing functionality is moved into the interposer itself (Chap. 4). That is, it is expected that TSI will migrate toward a full heterogeneous 3D implementation, with the Si interposer including I/Os, chip utilities (clocks, busses, networks), analog, or even electro-optic functionality. And interposers that include interconnect only will migrate toward lower cost implementations, such as POI or Hybrid solutions.

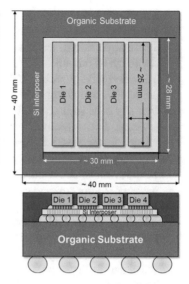

Large Split Die 2.5D SiP Technology Spec			
Item	I/O	D2D	P/G
Number of Interconnect Layers	3+ interposer / 8+ substrate		
Min Line/Space M1	10u	2u	N/A
Min Line/Space M2	10u		
Min Line/Space M3 and up	>10u		
Bump Pitch	80 -120u	40u	120u
Die Size (max)	Multiple 25mm x Xmm		
Package Size (max)	> 35mm x 35mm		
PoP	NO		
Lid & HeatSink	YES		

Fig. 2.33   Generic concept specification for the large split Die 2.5D SiP—a cartoon illustration of a concept Through-Si-Interposer (TSI) SiP construction, showing a cross-sectional and a top-down view of an Split Die FPGA package, and including a tabulation of typical target specifications

2.5.2   2.5D Technology Solutions for Area-Based Split Die SiP

Selection of appropriate 2.5D Technology Solutions for a Split Die SiP application, clearly depends on the application itself, and is mostly dominated by the total area of Si SoC that is being split.

  i. *Large Area Split Die*: The current, target specification for the 2.5D technology targeting *Large* Split Die SiP Implementation, such as for example used for high end FPGA products, is roughly as illustrated in Fig. 2.33.
     Given the die sizes and package dimensions involved, as well as some of the thermo-mechanical requirements, this application space is more like the HBM Memory Integration, and TSI is the 2.5D technology of choice. That is, in spite of the fact that the value proposition is (Large) Split Die, the manufacturing implementation is closer to the Memory Integration solution. Thus, the technology, the design tradeoffs, and the commercial implementations are similar, and the current commercial products are all based on TSI, or similar technologies. Note that Intel/Altera (Intel 2016) have announced an Embedded Bridge (EMIB) technology where the tight pitch D2D interconnect is contained in a Si sliver that underlaps the die to be connected, rather than covering the entire active die area. This approach clearly saves the cost of the Si

interposer, but requires complex process to embed this sliver into an organic substrate and to ensure the planarity necessary for 40 u uBump pitch attach. These applications, and the announced 2.5D technology solutions are also not the focus of this book.

The alternative possible 2.5D technology contenders for Large Split Die are either POI technology, or Hybrid TSV-Less Si Interposer + Substrate solution. As discussed above, in the long run it is believed that applications that leverage Si interposers that include interconnect wires only will migrate toward lower cost TSV-less implementations, and that Si based interposers will migrate towards higher value 3D implementations where the interposer includes much more than just the interconnect wires (I/O, utilities, analog, network, electro-optics, etc.). That is, in the long run, the (high) cost of Si Interposer will have to be justified by more than just interconnect wires, by including incremental higher value features.

ii. *Medium Area Split Die*: The current, common denominator specification for the 2.5D technology targeting *Medium* Split Die SiP Implementation are illustrated in Fig. 2.34:

This represents a reasonable "worst case envelope" of requirements targeting mobile type of potential applications, i.e., this is an aggressive set of target requirements which could satisfy most of the current high volume, commercial implementations of the split die value proposition, targeting mobile markets using the 14–10 nm FinFET CMOS generations of Si technology. The focus in this section is on the Medium sized Split Die for high volume commercial applications and not on the Large FPGA type of a Split Die implementation. Given this set of target requirements, it is not obvious which of the candidate 2.5D technologies would be the best solution. TSI is an unlikely contender due to cost and thickness constraints, but LCIs, LCIg, POI, FO WLP and Hybrid SLIM could all meet the fundamental implementation objectives, but with different tradeoffs. In principle, each of the candidates could be a viable option for Split-Die SiP (as defined by the spec above), with some general caveats listed below:

- LCIs is not as easy as may be thought, and there are several technical challenges. Its cost will probably limit its use to opportunities with higher margins than is typical for mobile or wearable market. In the long run, it is believed that TSV-less solutions will dominate. Nevertheless, the use of a Si interposer brings many advantages and, currently, it is a lower risk solution than other technologies.
- POI solution requires more development effort as integration of High Density layers in standard substrate process is hard. The effort required to make it work is probably justified only for applications that leverage other substrate value propositions such as x-y size or BLR considerations—i.e., an unlikely candidate for a Medium Area Split Die value proposition.

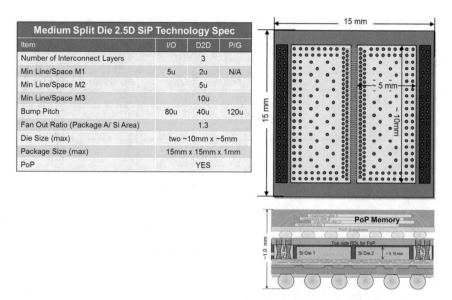

Medium Split Die 2.5D SiP Technology Spec			
Item	I/O	D2D	P/G
Number of Interconnect Layers		3	
Min Line/Space M1	5u	2u	N/A
Min Line/Space M2		5u	
Min Line/Space M3		10u	
Bump Pitch	80u	40u	120u
Fan Out Ratio (Package A/ Si Area)		1.3	
Die Size (max)		two ~10mm x ~5mm	
Package Size (max)		15mm x 15mm x 1mm	
PoP		YES	

Fig. 2.34 Generic concept specification for the medium split Die 2.5D SiP—a cartoon illustration of a concept SiP construction based on FOWLP technology, showing a cross-sectional and a top-down view of an Split Die SoC package, and including a tabulation of typical target specifications

- LCIg solution is attractive in principle, but both, the technology challenges and the supply chain considerations, inhibit its application for Medium Split Die SiP in the short run. In the long run it is believed that this technology will be optimized for rf applications due to the intrinsically attractive dielectric properties of glass.
- FO WLP is an attractive candidate especially for applications that favor very thin implementation and that do not require interconnect density lower than ~ 5 um line/space spec. However, it is believed that in the long run, the cost structure will be competitive only if the process adopts panel format.
- Hybrid SLIM is very attractive candidate especially for Split Die implementations that require higher interconnect density ($< \sim$ 2–5 um line/space) and have spatial constraints—such as mobile and/or wearable IoT devices. The cost structure is slightly higher than FO WLP, but the technology is scalable, and will benefit from additional demand from applications that leverage all other small die value propositions.

Table 2.3 below summarizes the *current subjective* ranking of each of the candidates versus the Medium Split Die SiP Value Proposition, with the spec as defined above, for various key attributes (1 is best, 5 is worst).

Table 2.3 Ranking of technology candidates for split Die SiP value proposition—a tabulation of a subjectively assigned rank of each of the viable 2.5D technology candidate for selected attributes relevant for Medium Area Split Die Value Proposition, on a of relative scale (1 is best, 5 is worst)

	LCIs	POI	LCIg	FO WLP	SLIM
Engineering risk	1	3	5	2	4
Manufacturability	1	3	5	2	4
Form factor (z)	3	5	4	1	2
Reliability CPI	1	2	3	4	5
Reliability BLR	5	1	4	2	3
Cost (per unit in HVM)	5	3	4	1	2
Cost (infrastructure)	2	4	5	3	1
Scalability	2	5	3	4	1
Supply chain	4	1	5	2	3
Overall ranking	**2**	**4**	**5**	**1**	**3**

Note that the overall ranking is arbitrary unweighted numerical summation of the individual rankings but that in reality, any product application will favor some attributes over others, i.e., in reality the overall ranking will be different for different applications. For mobile and IoT wearable space, for example, the thickness and cost considerations could trump other attributes, and FO WLP and SLIM solutions would be the favored contenders.

iii. *Small Area Split-Die*: as indicated in Sect. 2.3.3, the area-based Split Die value proposition can be realized at quite small die sizes, mostly due to high cost per sq mm of Si associated with very advanced CMOS nodes. Note that it is believed that the power-performance requirements associated with Mobile and IoT Wearable markets will favor moving to most advanced CMOS nodes, necessary to capture this value proposition. However, this value proposition can be realized only with very high density D2D interconnect—below ∼2 um line/space spec. The basic organic materials typically used in classical OSAT and Substrate packaging technologies do not have the dimensional stability commensurate with this specification range. Hence POI, FOWLP, and Glass Core Substrate LCIg, technologies are unlikely candidates for this application space. Si-based technologies,i.e., LCIs and Hybrid SLIM are the likely solutions.

It is believed that in the long run, the likeliest candidate is the TSV-less Hybrid SLIM type of a technology—mostly due to the cost advantage over the fully processed Si Interposer solution. This approach does not call for major investment in upgrading the OSAT infrastructure, and preserves the interconnect scalability necessary for small area Split Die. In addition, Hybrid SLIM technology offers the form factor advantages which are important for mobile and IoT Wearable space.

Table 2.4 Ranking of technology candidates for small Die value proposition—a tabulation of a selected key attributes of the two best candidate technologies for the Small Die Value Proposition

	LCIs	SLIM
Starting Si wafer cost	Semiconductor grade wafers ($80 to $100)	May use test wafer grade material ($10–$20)
Foundry process	Dual-Damascene M1 + TSV masking and processing (TSV etch, liner, fill)	Dual-Damascene M1 + Etch-Stop Dielectric film on bottom
OSAT process	TSV Reveal processing (Grind, CMP polish, dielectric, mask, pad)	Grind and Si Bulk Etch
Warpage and Planarity (in line and end of line)	Leverage Si itself as a carrier for part of flow. No CTE mismatch versus Si die	Requires (careful) carrier management. CTE mismatch versus die
Reliability (CTE match versus PCB)	BLR concerns with the interposer at risk. Limited package size	CPI concerns with Si die at risk. Limited package size
OSAT Assembly Flow	Typically Die-First—leverage planarity advantage of Si. Tight uBump pitch	Can do die-first or die-last, with tradeoffs of cost versus uBump pitch
Cost (per unit in HVM)	Relatively high—estimated at about ∼50% higher than SLIM	Similar to wafer based FO WLP cost structure
Cost (infrastructure)	Some Investment Required—to enable OSATs to do MEOL processing	Very Low—good distribution of requirements versus supply chain

2.5.3 2.5D Technology Solutions for Small Die SiP

As is the case for Small Area Split-Die SiP, the technology requirements for the non-area based value propositions tend to be associated with small die (PHY die and/or Tiny Mono-Die). Therefore, again, the technology options based on conventional organic materials—POI, FOWLP, LCIg—are not realistic candidates. Technologies based on Si Interposer, that leverage foundry infrastructure, offer scalable interconnect pitch required for accessing the very small die, i.e., LCIs and Hybrid SLIM are the realistic contenders. SLIM cost structure is more like FOWLP and is more attractive than that of LCIs. In addition, SLIM preserves the Form Factor advantages favored by the Mobile and IoT Wearable markets. On the other hand, LCIs is lower risk and more scalable flow.

Table 2.4 summarizes some of the differentiating factors and attributes:

Therefore, it is believed that in the long run, Hybrid SLIM type of technology will dominate this space, due to its good match to the requirements of mobile and IoT Wearable markets.

References

AGC (2016) EN-A1 glass for TGV, WLP and MEMS. http://www.agcem.com/index.php/advanced-packaging/3-en-a1-glass-for-tgv-wlp-and-mems. Accessed 24 Nov 2016

AMD (2016) High Bandwidth Memory, Reinventing Memory Technology. http://www.amd.com/en-us/innovations/software-technologies/hbm. Accessed 24 Nov 2016

Black B (2012) Sie stacking and the system. In: IEEE hot chips HC24, Aug 2012

Chen YH et al (2014) Low cost glass interposer development. In: International symposium on microelectronics, vol 2014, No 1

Corning (2016) Corning semiconductor glass wafer products. https://www.corning.com/worldwide/en/products/advanced-optics/product-materials/semiconductor-laser-optic-components/semiconductor-glass-wafers.html. Accessed 24 Nov 2016

eSilicon (2016) Custom 2.5D & 3D Packaging/ HBM Solutions. https://www.esilicon.com/services-products/services/custom-2-5d-3d-packaging/. Accessed 24 Nov 2016

Gargini P (2016) Roadmap—Past, Present and Future. https://www.spcc2016.com/wp-content/uploads/2016/04/02-01-Gargini-ITRS-2.0-2.pdf. Accessed 24 Nov 2016

Huang T et al (2014). Adhesion and reliability of direct Cu metallization of through-package vias in glass interposers. In: IEEE 64th electronic components and technology conference (ECTC)

Intel (2016) Embedded Multi-die Interconnect Bridge. http://www.intel.com/content/www/us/en/foundry/emib.html. Accessed 24 Nov 2016

International Technology Roadmap for Semiconductors 2.0, 2015 Edition (2015). http://www.itrs2.net

International Technology Roadmap for Semiconductors, 2005 Edition (2005). http://www.itrs2.net/itrs-reports.html

Ivankovic A et al (2015) 2.5D Interposers and advanced organic substrates landscape: technology and market trends. In: International symposium on microelectronics, vol 2015, No 1

Kang T, Yee A (2013) A comparison of low cost interposer technologies. In: IEEE CPMT Society. http://www.ewh.ieee.org/soc/cpmt/presentations/cpmt1305d.pdf. Accessed 24 Nov 2016

Kelly M et al (2016) System level IC packaging—integration of key technologies: TSVs, interposers, and advanced IC packaging. In: iMAPS additional conferences (Device Packaging, HiTEC, HiTEN, & CICMT): January 2016, vol 2016, No. DPC

Kelly M et al (2015) 2.5D and 3D Multi-die packages using silicon-less interposers. In: iMAPS 11th international conference and exhibition on device packaging fountain hills, Arizona USA, 17-19 Mar 2015

Kim J, Kim Y (2014) HBM: memory solution for bandwidth-hungry processors. In:IEEE hot chips, HC 26 Aug 2014

Kiyoshi Oi et al (2014) Development of new 2.5D package with novel integrated organic interposer substrate with ultra-fine wiring and high density bumps. In: 2014 IEEE 64th Electronic components and technology conference (ECTC)

Kwon WS et al (2014) Cost effective and high performance 28 nm FPGA with new disruptive silicon-less interconnect technology (SLIT). In: International symposium on microelectronics, vol 2014, No 1

Leachman RC (2002) Competitive Semiconductor Manufacturing: Final Report on Findings from Benchmarking sub-350 nm Wafer Fabrication Lines, University of California at Berkeley. http://www.microlab.berkeley.edu/csm/csmreports.html. Accessed 24 Nov 2016

Lee C-C, Black B et al (2016) An overview of the development of a gpu with integrated HBM on silicon interposer. In: IEEE 66th electronic components and technology conference (ECTC)

McCann S et al (2014) Flip-chip on glass (FCOG) package for low warpage. In: IEEE 64TH electronic components and technology conference (ECTC)

Remi Yu (2012) Foundry TSV enablement for 2.5D/3D chip stacking. In: IEEE hot chips HC24, Aug 2012

Ryuta F et al (2015) Demonstration of 2 μm RDL wiring using dry film photoresists and 5 μm RDL via by projection lithography for low-cost 2.5D panel-based glass and organic interposers. In: IEEE 65th electronic components and technology conference (ECTC)

StatsChipPac (2016) Innovative Fan-out Wafer Level Technology. http://www.statschippac.com/services/packagingservices/waferlevelproducts/ewlb.aspx. Accessed 24 Nov 2016

Sukumaran V et al (2012) Low-Cost thin glass interposers as a superior alternative to silicon and organic interposers for packaging of 3-D ICs. In: IEEE transactions on components, packaging and manufacturing technology, vol 2, Issue 9

Tseng CF et al (2016) InFO (Wafer Level Integrated Fan-Out) technology. In: IEEE 66th electronic components and technology conference (ECTC)

tsmc (2016) CoWoS® Services. http://www.tsmc.com/english/dedicatedFoundry/services/cowos.htm. Accessed 24 Nov 2016

Xilinx (2016) The Industry's Only All Programmable Homogeneous and Heterogeneous 3D ICs. https://www.xilinx.com/products/silicon-devices/3dic.html. Accessed 24 Nov 2016

Yoon SW et al (2013) Fanout flipchip eWLB (embedded Wafer Level Ball Grid Array) technology as 2.5D packaging solutions. In: IEEE 63rd electronic components and technology conference (ECTC)

Yu D (2015) A new integration technology platform: Integrated fan-out wafer-level-packaging for mobile applications. In: IEEE/ JSAP 2015 symposium on VLSI technology (VLSI Tech)

Chapter 3
More-than-Moore Technology Opportunities: 3D SiP

3.1 Overview

As discussed in Chap. 2 there are two generic types of SiP technologies:

i. multiple die in a single package placed side-by side. This is sometimes referred to as "2.5D Integration"
ii. multiple die in a single package stacked on top of each other using Through-Si-Vias. This is referred to as "3D Integration"

The focus of this section is 3D Integration. Integration in the third dimension is clearly perceived by futurologists and technologists alike, as the ultimate path toward continued system-level integration. After all, human brain—the ultimate processing machine—is clearly a 3D integrated system. The advantage of 3D integration is obvious and, at the highest level, boils down to the ability to interconnect lots of logic with shortest wires, which in turn precipitates other derivative value propositions, such as highly parallel interface, lower power, higher performance, smaller form factor, etc. Clearly, the ability to leverage this value proposition is directly related to the number of Tier-to-Tier (T2T) connections, which, in turn, is related to the size and density of the TSVs (and uBumps).

There are several ways of stacking multiple die on top of each other, i.e., there are multiple classes of 3D technologies and 3D SiPs. For example, PoP memory scheme—ubiquitously used in mobile devices—is strictly speaking a 3D technology, that stacks a memory package on top of a logic package. Similarly, integration schemes that stack bare memory die, rather than packaged die, on top of logic, using, for example, wire bond (Amkor 2016; ASE 2016; Fujitsu 2016) or FO WLP (Merritt 2016; Braun 2016; StatsChipPac eWLB 2016) type of technologies are also 3D SiPs. Similar 3D schemes have also been used for integration of analog, or other types of die in a multi-package component.

© Springer International Publishing AG 2017
R. Radojcic, *More-than-Moore 2.5D and 3D SiP Integration*,
DOI 10.1007/978-3-319-52548-8_3

This chapter focuses on one class of 3D Technologies; namely those using Through-Si-Vias (TSV) interconnect. The reason for this is that TSV-based 3D technology enables high density, nonperipheral, interconnect, with 1000s of wires between the die, as opposed to the 10s or 100s of pins, with peripheral I/Os, enabled by PoP or folded FO WLP technologies. As such, TSV-based 3D technology is seen truly as an *integration* technology opportunity, rather than an extension of existing packaging technologies. Effective use of TSV-based 3D technology impacts the entire product ecosystem, driving changes in manufacturing (Si and packaging process technologies) and design (architecture and chip design) paradigms, and as such is a disruptive 'More-than-Moore' type of a technology opportunity. Thus, whereas PoP and FO WLP technologies are flavors of 3D assembly and packaging technologies used within the More-Moore paradigms, TSV-based 3D die stacking represents a shift toward More-than-Moore paradigm.

Note that the term "2.5D Technology" used in Chap. 2 was coined to describe a SiP package solution that is in-between the current planar Mono-Die ("2D") SoC, and the stacked die (3D) solution, with some features in common—namely the TSV.

A number of different TSV technologies have been discussed in the industry with potential application ranging from, for example, stacked CMOS Image Sensors for compact cameras, to stacked CPUs for high performance digital processing. Consequently, TSV dimensions ranging from the very coarse (many 10s of um in diameter), to very fine (in the sub um dimensions approaching those of standard BEOL vias) have been proposed. The focus here is on TSVs of the order of few microns in diameter—as required for die stacking for 3D SiP solutions with the number of tier-to-tier connections typically in the 1000–10,000 range. This type of TSV technology is the current state of the art and is ramping up in high volume

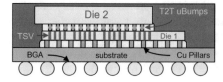

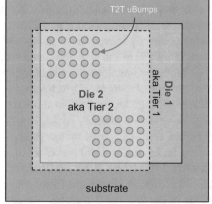

Fig. 3.1 Concept generic 3D SiP package—a cartoon illustration of a generic concept 3D SiP Package, showing cross sectional and top down views, and highlighting the major differentiating features (Through Si Vias (TSV), multiple die, uBump Tier-to-Tier attach), and identifying some standard package features (Cu pillars, BGA, substrate, mold)

manufacturing in real commercial products, i.e., this class of TSV's is a reasonable, practical, and realistic technology for use in commercial SiP products in the foreseeable future.

The sketch in Fig. 3.1 illustrates the general 3D Stacked concept considered here, and identifies the following key differentiating features:

i. *Through Si Via* (TSV)—basically a via providing a connection from the front to the back of a die
ii. *Tier to Tier* (T2T) *uBump*—basically a (small) pillar connecting the two die
iii. *Multiple Die* on a single substrate package, along with usual packaging features (Cu Pillars or C4, underfill, mold, BGA etc.).

3.2 3D SiP Technology Options

Unlike the case of 2.5D Integration technology that is characterized by diverse set of baseline technologies 3D Integration technology addressed here revolves around a single central feature: the Through Si Via (TSV). That is, the 3D Integration options discussed here all leverage the TSV technology, which offers a rich diversity of integration flows, but not a diversity of baseline technologies (e.g., substrate FCBGA, FO WLP, Si Interposer, etc.). Consequently, this section focuses on TSV formation and Integration options.

3.2.1 Requirements

The key differentiating attribute of TSV-based 3D technologies is to enable high density ($\gg$1000) of inter-die connections—which currently cannot be commercially achieved by any other technology.

One of the early drivers (circa 2008–2012) of 3D TSV-based technology in digital SIP domain was Memory-on-Logic stack, leveraging the value propositions derived from WideIO memory (Sect. 3.3.3). In that case, the T2T interface requirements were defined by the JEDEC standards (JEDEC 2011, 2013, 2014), and the DRAM memory vendors (Samsung, Hynix, Elpida, etc.). Ergo, the early industry efforts focused on TSV technology candidates that could enable an interface between a DRAM Memory and a Logic SoC die and required $\sim$1000+ connections.

Given the various trade-offs associated with the 3D options, (Sect. 3.4) and specifically for the TSV technology (area overhead, wafer thickness, etc.), the common requirement for the TSV-based 3D chip stacking technology is as following:

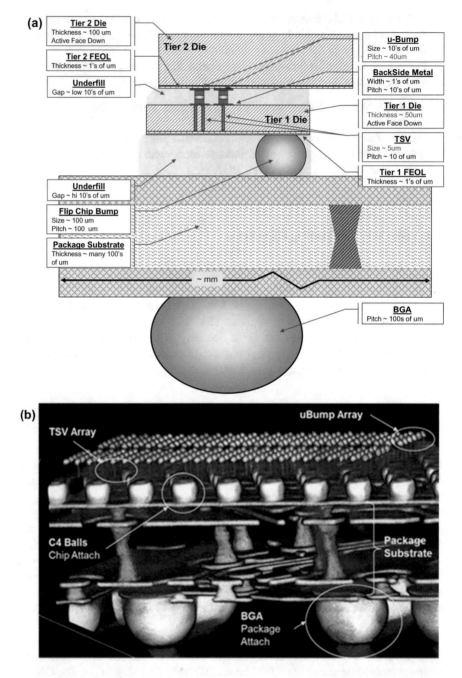

Fig. 3.2 Generic TSV based 3D SiP package—a cartoon illustration of a generic concept TSV-based 3D SiP Package showing a two die face-to-back (F2B) stack and highlighting the differentiating (TSV and the T2T uBump) and standard (C4 chip bumps, BGA balls, substrate, underfil, etc.) features. Also included is an X-ray tomography photograph highlighting the relative sizes of these key features

- TSV diameter: ~5 um
- TSV depth: ~50 um
- TSV fill: Cu
- uBump pitch: ~40 m
- uBump diameter: ~25 um

These requirements—often referred to as the 5/50 TSV—are illustrated in the sketch and the photograph in Fig. 3.2, and represent a sort of "center of gravity" of the industry effort with 3D stacking. The illustration is approximately to scale for the key features identified. The configuration shown is Face-to-Back (F2B) with both Tier 1 (logic) and Tier 2 (memory) dice active side down, with uBump T2T attach. The chip-to-substrate attach shown in the sketch is a C4 solder ball, but Cu pillar could be (and often is) used.

Different applications dictate different requirements for the TSV—such as the 10 um/100 um TSV typically used for Si Interposers, or the ~<1 um/2 um TSVs used for 3D memory implementation (e.g., Tezzaron 2016). However, the 5/50 TSV, or implementations in that approximate range, with TSV diameter varying between ~3 and ~8 um has been demonstrated in volume manufacturing, and is an excellent vehicle to explore the various technology and design trade-offs associated with 3D stacking.

3.2.2   TSV Formation

TSV concept—i.e., forming a connection from the front to the back of a Si die—and the basic associated technologies are not particularly new. The original patents have been filed five decades ago (Smith and Stern 1964), and there are multiple technology flows that have been proposed and pursued to "drill and fill" a TSV. The fundamental process technologies for formation of the TSV ("drill"), isolation and filling of the TSV ("fill"), the thinning and capping of the bottom of the TSV ("reveal"), and the integration of these modules into the overall flow are briefly described here, in order to recap and summarize, rather than to provide new insights.

The key steps for the TSV (Via-middle) formation are as follows. Note that the process for other flavors of TSVs (via-first, via-last), shares many similar steps, but use different insertion point and/or different materials:

- **Etch**: almost all variants for formation of the TSV use the so-called Bosch Process consisting of a successive series of plasma etch and clean steps. This is required to prevent deformation of the via due to re-deposition of the etched material. Consequently, the TSV walls typically end up being "scalloped" at a microscopic scale, as shown in Fig. 3.3. The throughput (and hence the cost) of this process has to be balanced versus the overall quality and yield, and specifically the ease of forming a good liner and fill on the sidewalls and bottom of a deep via.

Fig. 3.3 Through Si Via—a
Micrograph (Chipworks) of
portions of a TSV,
highlighting the key attributes
of the etch (sidewall
scalloping), liner (coverage at
the bottom of the blind via),
and fill (Cu with no voids)
process steps

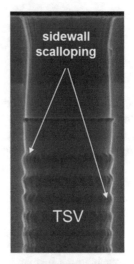

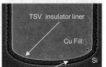

- *Liner*: the TSV clearly needs to be isolated from the semiconducting Si bulk—
  and hencethe via needs to be lined with a dielectric. Clearly the integrity of this
  liner is essential to managing TSV leakage currents and reliability. Forming a
  continuous dielectric film, with no cracks or creases, and with good thickness
  uniformity, at the sidewalls and bottom of a deep via is much harder than on a
  clean planar surface. The TSV forms a cylindrical M-O-S capacitor, so that the
  via capacitance is not only controlled by the liner thickness, but also by the Si
  substrate doping levels and the nature and density of the interface states.
  Controlling the interface in a deep via is challenging especially because some of
  the characteristics may be further affected by the reveal processing further down
  in the flow (thinning, polishing, passivating the backside)
- *Fill*: the most common fill metals used are Cu and W, each with its own
  constraints and advantages. Cu has better electrical performance, is ubiquitously
  used in IC and packaging industries, and is easily plated. But it is relatively soft
  and mobile, has very different CTE than Si, and cannot sustain high temperature
  ($> \sim 400$ C) processing. W is compatible with high temperature processing and
  is very tough and stable, but is quite resistive and has a number of processing
  idiosyncrasies. Whereas it is not susceptible to creep, like Cu, it suffers from
  different stress challenges. Cu is typically used for via-middle and via-last TSV
  flavors, albeit some specific implementations favor the use of W to avoid Cu
  stress management issues. TSV Cu fill is a two-step process, where a seed layer
  (usually Ti/TiN) is first deposited, and then the via is filled by Cu electroplating.

Fig. 3.4 TSV based 3D die stack—a micrograph of a section (TechInsights) of a multi-die stack, showing the TSV and the T2T uBump, and highlighting the (∼10:1) aspect ratio of the TSV, and the relative dimensions of the key features

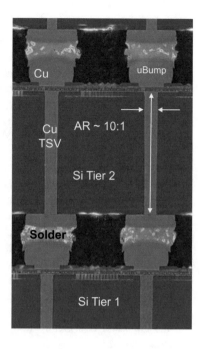

Note that the plating rate (and hence cost and quality) needs to be managed carefully to prevent closing off the neck of the via resulting in a void in the via center (aka key-holing)

- *Polish and Cap*: the top of the plated via fill is usually domed or mushroomed, so that the via formation flow is completed with a CMP polishing step to planarize the via, and a dielectric film deposition to cap the top of the TSV.
- *Reveal*: TSVs are drilled as blind vias into full thickness wafer, and upon completion of the TSV and rest of the CMOS processing, the wafer is flipped and thinned to reveal the bottom of the TSV. The reveal process is usually a coarse Si grind followed by a CMP polish step, in order to prevent excessive dislocation or other imperfections nucleating on the backside surface. The reveal process is completed by depositing and patterning a dielectric layer to define the backside pad and/or RDL layer and passivate the backside surface as illustrated in Fig. 3.4.

3.2.3   TSV Integration with CMOS Flow

Integration of the TSV with the rest of the CMOS processing is a separate set of challenges and constraints, and several integration schemes have been proposed and are pursued by the industry. These are typically classified by the insertion point in the standard CMOS flow, as following:

- *Via-First*: The TSV deep etch, and the associated via liner and fill steps, is implemented during standard CMOS FEOL processing—typically around steps used to define the Shallow Trench Isolation (STI). This flow leverages well the existing Si technologies, and can realize the highest aspect ratios (>10:1), and hence the smallest vias, but requires the materials to be tolerant to the subsequent FEOL process temperatures needed to define the CMOS devices. This, in practice dictates the use of W, or similar refractory metals, as the fill material.
- *Via-Last*: The TSV etch, line and fill steps, are implemented after completion of the all of Si processing—thereby removing the requirement for tolerance to high temperature processes, and readily allowing use of Cu as the fill material. The TSV is drilled from the back of the wafer, after thinning to the desired thickness. Currently, this approach is typically implemented to land the TSV on higher levels of metal, or even on the bottom of the pads on front of the wafer. Given the front-to-back alignment challenges, as well as issues of drilling through FEOL and BEOL materials, this class of vias is typically fairly low aspect ratio (<5:1) and is fairly large. In addition, via-last approach demands keep out areas for both FEOL and BEOL features, and is as such expensive in terms of area utilization. On the other hand, via-last technology avoids some integration and scaling challenges faced by via-middle technologies. Altogether, this approach is typically not used for applications that call for high density interconnect.
- *Via-Middle*: with this flow the TSV deep etch, line and fill, is implemented after the completion of the FEOL Si processing, but before the BEOL processing. This flow leverages Si processes, and can realize high aspect ratios (~ 10:1), but also does not require tolerance to the high FEOL process temperatures—thereby allowing use of Cu as the fill material. Cu filled via-middle type of a TSV (5/50 or 10/100) has become the most ubiquitous, especially for digital applications (memory cube stacking, interposers, etc.). Integration of the relatively large TSV features (um's) with the processing of the poly, contact, and metal layers—all features in the nm range—is challenging. Hence via-middle technology is progressively harder to integrate with scaled CMOS technologies. On the other hand, via-middle is only a placement blockage, requiring keep out area for FEOL features only, and allowing unconstrained routing over it, so that it is less expensive in terms of area overhead, than via-last.
- *Backside Via*: Recently, a fine-grain version of via-last TSV technology, with diameter of $\sim$ few um, has been proposed. That is, this is a version of via-last approach but it lands the TSV on the local interconnect or Metal-1, rather than drilling through the BEOL stack. This via type therefore can be as area efficient as via-middle. Modern alignment technologies, without the need to image through the BEOL stack, enable fairly accurate placement and since this is end-of-line process, Cu fill can be used. In addition, in principle this process could be implemented in the OSAT line, thereby offering foundry independent flow and leveraging the cost advantages of the OSAT. Furthermore, with this technology there is an opportunity to eliminate some of the steps for backside processing—by forming a Cu stump as a part of the TSV plating, and

eliminating the need for masked bump pad layer. However, the technology for this type of a via is in development now, and there is no demonstration of manufacturability, as yet. Research entities (IMEC 2016; IME 2016) and some equipment vendors (PTI 2016) have been exploring this option.

- *Monolithic TSV*: The ultimate 3D stacking technology—so-called monolithic TSV is to realize a multi-tier sandwich of active Si with BEOL interconnect. This is normally done by leveraging Silicon-on-Insulator (SOI) and/or epitaxial deposition technologies, where a layer Si is grown on top of the BEOL stack of a completed wafer, and is then processed to realize another layer of active devices. Since the layers can be made very thin, the tier-to-tier connections are realized by a very small via connection using standard BEOL via processes. The drawback is obviously that the active devices in first tier must withstand the processing temperatures associated with building the active devices in the second tier—i.e., managing the dT during the processing may be challenging. This type of technology is pursued by MIT labs, LETI, Monolithic (MIT Lincoln Lab 2016; LETI 2016; Monolithic 3D 2016).

- *Hybrid Processes*: several unique flavors of TSV technology have also been deployed. Some examples: (a) Elpida (now Micron) and PTI have developed a hybrid TSV flow that combines via-first and via-last, where the TSV is defined during STI steps, but the (Cu) fill is implemented in an OSAT after completion of all the Si processing; (b) Tezzaron has a specialty TSV technology based on a hybrid (Cu-Cu/O-O) wafer-to-wafer (W2W) bonding, thinning, and realizing the T2T interconnect with tiny (<1 um) W filled vias (they call them super contacts). These approaches sidesteps the thermal budget management challenges of Monolithic technologies, but offers comparable T2T density.

3.2.4   TSV Integration with Assembly Flow

In addition to integration with the rest of CMOS processing in the Si fab line, there are further challenges of integrating Si wafers with TSV's into the packaging flow. These are summarized below:

i. *Post Wafer Fab Flow*: The TSV Reveal processing (thinning, polishing, backside prep) can be done either by the Si foundry or an OSAT.
  The advantage of using the foundry is that a single entity is then responsible for building and testing the completed wafer with TSVs, thereby presumably enabling better opportunity for learning and yield improvement. However, this would then necessitate shipping thin (e.g., 50 um), double-sided wafers—which is possible but awkward.
  The advantage of performing the TSV Reveal Processing at the OSAT is that it can be integrated into the assembly flow, with opportunities to leverage

standard OSAT processes steps—e.g., bumping , carrier bond/de-bond, etc. This is sometimes also called "Middle End of Line" (MEOL) flow—a misnomer that has stuck in the industry—especially for Si interposer processing (Chap. 2). Note that OSATS favor use of thick organic (PI) dielectrics for backside dielectric whereas the foundries have ready access to thin inorganic dielectrics (Si oxide and/or nitride). This choice has implications for warpage management, stacking sequence, cost, quality, etc.

ii. **Stacking Flow**: definition of the stacking flow is a key knob in the 3D assembly flow that has major impact on the overall architecture and total cost. The options are the following:

- Wafer-to-Wafer (W2W)
- Die-to-Wafer (D2W)
- Die-to-Die (D2D)
- Die-to-Substrate (D2S)

W2W flow is the most attractive from the cost point of view, since it is gang processing rather than die-by-die processing. However, it also imposes limitations on the die sizes (the two die must be the same size) and die yields (the yields must be very high to avoid stacking good die on bad die). Hence this is a viable option mostly for homogenous memory stacking—since the die is the same size, and memories typically include redundancy and other repair schemes that boost die yield to acceptable levels.

D2W process is also attractive since it uses one of the wafers as a carrier. However, this attach scheme requires that T2 die be equal to or smaller than T1 die, and that it is stacked within the footprint of T1 die, i.e., no overlap/underlap (Sect. 3.4.3) stacking is possible.

Die-to-Die (D2D) and Die-to-Substrate (D2S) flows are one-die-at-a-time flows, and are hence expensive. They pose constraints on the assembly process and dictate different sequence and number of temporary bond/debond steps, die attach process (thermal compression vs reflow), and so on—all of which have impact on the cost of packaging. On the other hand, these two flow types are less restrictive with relative die sizes and placements and offer maximum amount of flexibility in design and architecture.

3.3   3D SiP Technology Value Propositions

At the highest level of abstraction, the value proposition for 3D stacking is similar as for the 2.5D side-by-side integration and is the usual 'short wires' (and hence good Power-Performance) and 'form factor' propositions. In general, in addition to the obvious form factor benefits, when the inter-die interconnect requires much more than 1000–2000 wires, 3D integration may be more advantageous than 2.5D integration, due to the overheads associated with each integration scheme. However, at the next level of detail, the situation is complex and involves alternative avenues for system level integration, with many cross-functional trade-offs. These are summarized below:

3.3.1   Homogenous 3D SiP Integration Value Proposition

Homogenous 3D Integration involves stacking similar—or identical—die on top of each other, and the value proposition is enhanced density of gates and/or bits. Specifically

i. **Memory-on-Memory (M-o-M)**: TSV-based 3D die stacking technology is today in Volume Manufacturing—for implementation of 3D memories. All major memory manufacturers have one, or more, products that leverage TSV technology for stacking memory die into memory "cubes." In addition, some architectures also leverage TSV-based technologies for integration of the memory cubes with the logic controllers, as summarized in the Table 3.1.

   High Bandwidth Memory (HBM) (Hynix 2016) cube, targeting graphics, and HPC markets with 2.5D Interposer integration is qualified and in production. Hybrid Memory Cube (HMC) (Micron 2016), targeting high end server applications, has also had product announcements, and is presumably ramping into volume production. The 3DS DDR (Samsung 2016) memory is a higher performance/lower power (and higher price) replacement for DIMM cards, likely targeting tablets and PC market, and leveraging the existing DDR architecture (i.e., no major redesign). WideIO memory, targeting mobile market, and Memory-on-Logic 3D die stacking, is one 3D M-o-M/M-o-L technology that did not take off. JEDEC standards describing these memory architectures are available (JEDEC 2011, 2013, 2014). Hybrid Memory Cube, High Bandwidth Memory, and WideIO schemes all involve stacking 2, 4, or 8 DRAM die (currently 1 GB each) using TSV (usually via-mid) and uBump

Table 3.1   3D DRAM memory solutions—a summary of the various commercial 3D TSV-based DRAM memory solutions, including their respective target applications, status, and key attributes. Also highlighted is the technology intended for integration with the processor die

	3DS DDR4	HMC	HBM	WideIO 1	WideIO 2
Market	PC	HPC	HPG/HPC	Mobile	Mobile
Spec	JEDEC December 2013	HMCC 2014	JEDEC October 2013	JEDEC December 2011	JEDEC August 2014
HVM	YES	Ramping	YES	NO	NO
Interface	32/64	SerDes	1024	512	256/512
Integration	PCB based 2D	PCB based 2D	TSV based 2.5D	TSV based 3D	TSV based 3D

technology to interconnect them, into a memory "cube." For HBM and HMC solutions the cube is then stacked on top of a specialized memory-manager die, again using TSV/uBump interconnect, and assembled into a single package. HMC targets low-latency, high performance computing applications, and uses a high speed SerDes for communication with the CPU die. HBM is optimized for high bandwidth applications and uses a wide interface (1024 bits) to logic. WideIO is discussed below.

The core value proposition of 3D Memory-on-Memory stacks is derived from packing more bits per unit volume (and area) and thus achieving better density. This physical density advantage invariably translates to better core power-performance characteristics, i.e., it takes less energy to move a memory bit a shorter distance. Use of parallel architectures and advanced memory-logic integration schemes (either 2.5D or 3D) provides further and incremental I/O power-performance benefits.

Stacked M-o-M memory cubes are now commercially available and ramping into volume production, and are hence not the focus of this book. As is often the case in the semiconductor industry, memories paved the way, and have served very well to de-risk the 3D stacking technologies (both actually and psychologically), making the use of TSVs much more palatable in all other applications. It is in fact natural that memory products should be the first to adopt 3D die stacking: they are identical die that can leverage wafer-to-wafer (W2W) stacking to contain process costs, and an array design that includes redundancy and repair schemes to manage the yield loss associated with W2W stacking.

ii. *Logic-on-Logic* (L-o-L): The core value proposition of homogenous Logic-on-Logic stacking (both die implemented in same technology node) is shorter wire lengths. It is a matter of simple geometry to demonstrate that as 2D die size grows, it is possible to implement shorter interconnect by splitting the die and stacking two smaller die on top of each other. This value proposition can be realized only if the TSV and uBump pitches are commensurately tight, and there are no placement restrictions on either of the die. In addition, 3D L-o-L Die Stacking has the same potential Split-Die benefits as described above in Sect. 2.3.2 for 2.5D SiP. However, the trade-offs required to optimize the 3D SiP are somewhat different, as described more fully below (Sect. 3.4). In general, the value proposition for homogenous 3D L-o-L stacking is maximized with very fine grain partitioning, high interconnect density and small TSVs and uBumps—all in order to enable almost seamless "sloshing" of logic content across the multiple tiers. The design and process technologies required to enable this type of integration for commercial applications is judged to be beyond a $\sim$5 year horizon.

Thus, the value proposition for the homogenous 3D die stacking is pretty clear and is currently leveraged in memory-on-memory stacked commercial products. The adoption of L-o-L homogenous stacks is gated by the availability and maturity of design and EDA technologies needed to realize suitable commercially viable products. In addition, the process technologies needed to realize the dense T2T interconnect required for optimized L-o-L stacks (e.g., hybrid D2W bonding, sub-um TSV?) are not available, yet.

3.3.2   Heterogeneous 3D SiP Integration Value

Heterogeneous 3D Integration involves stacking different die on top of each other, and the value proposition includes form factor (vs multi-package integration), power (vs serial interface), and cost (vs Mono-Die SoC or 2.5D SiP) benefits, depending on the specific integration scheme. Candidate integration schemes that have been publicized and corresponding Value Propositions include the following:

  i. **Sensor-on-Logic**: some sensors—most notably CMOS Image Sensors (CIS)—prefer a wide interface to logic, and use of TSV technology (typically TSV via-last flavor) has proven form factor + performance value proposition that is in HVM today. Most cameras in smartphones leverage some form of 3D stacking of CIS on Logic. Advanced CIS manufacturers (e.g., Sony 2012) are pursuing architectures that require finer grade TSV technology and are pursuing TSV-middle + D2D bonding schemes.
 ii. **Heterogeneous Logic-on-Logic and/or Analog-on-Logic**: integrating different logic circuitry—each manufactured in a technology optimized for its function—is clearly a possible integration scheme that offers potential cost and form factor value propositions. All heterogeneous Split-Die opportunities could potentially be realized in either 2.5D or 3D SiP—e.g., the PHY Die opportunity discussed in Sect. 2.3.3 could also be realized by stacking the small PHY die on a logic SoC. In general, 3D Split-Die makes more sense than 2.5D if the wire count exceeds ~ 2000 wires—such that the lower overheads associated with T2T stacking than those associated with D2D interface are leveraged. The core value propositions are the same Split-Die opportunities as described in Chap. 2.
iii. **Memory-on-Logic**: Tight coupling between the main system memory and the logic CPU is clearly a basic function of any computational system. This is typically implemented via standard DRAM DDR interface between two packaged components—for mobile applications typically using PoP integration. The interconnect density is limited by packaging and PCB technologies, and is hence forcing the use of a relatively narrow memory channel interface—typically x16, x32, or x64 bits wide—as described by the various generations of the JEDEC DDR standards. A wide memory interface—e.g., 512 or even 1024 bits wide—can be enabled by the 3D stacking technology, with the incremental value propositions of reduction of power (for a given bandwidth), as well as smaller form factor (vs two separate packages or vs PoP). This is analogous to the 2.5D Memory Integration value proposition, using HBM memories and Si Interposers. A specific opportunity, architected for use with 3D TSV-based stacking is WideIO DRAM memory—see Sect. 3.3.3.

Tight coupling with memories other than the main system memory is also often desirable. Some applications—for example, Modems—benefit from a small "private" memory (typically ≪1 GB DRAM memory die). Use of TSV technology is not appropriate, and "integrated Stacked Memory' (ISM) is typically implemented by wire bonding a small separate DRAM die on top of the modem die or package. Similarly, a nonvolatile memory (Flash or maybe MRAM die) could be integrated with the logic SoC—using either PoP or wire-bond technology. TSV technology could be used to create connections on the back of the SoC die for memory attachment, to enable either a wider interface, or a very small form factor, or both. Whereas these applications appear to be very attractive—at least on paper—none have been implemented in a commercial product (as yet?), probably due to the cost and other trade-offs.

3.3.3   *WideIO Memory Value Proposition*

Wide IO DRAM memory has been architected specifically for use with 3D TSV-based stacking on a logic SoC die, especially targeting mobile market. Even though it has not been a commercial success, it has been evaluated extensively by all the principal participants in the mobile market at the time (West et al. 2012; Kim et al. 2013; Gu et al. 2012; Farcy et al. 2013; Samsung 2013) and as such is an excellent vehicle for describing 3D die stacking technology and the associated trade-offs. Hence WideIO and specifically WideIO Memory-on-Logic stacking is addressed here.

The core WideIO Memory on Logic (M-o-L) value proposition is derived from using wide memory interface driving small loads—resulting in a superior memory power efficiency. The TSV-based 3D stacking is the integration technology that enables this value proposition, i.e., the core value proposition is the wide interface, rather than the TSV per se. The industry developed WideIO1 and WideIO2 memories for implementation with TSV technology targeting mobile market (analogous as the HBM memory leverages TSV Si Interposers and targets GPU market). These products appear to have missed the market window of opportunity, and seem to have been supplanted by the incumbent LPDDR3 and LPDDR4 PoP implementations (so far). It is believed that if the mobile market demand for bandwidth continues to increase (potentially driven by bigger screens with better resolution), as it has in the past—see Fig. 3.5a—the industry shall revisit the trade-offs between PoP/LPDDRx and 3D/WideIO architectures. That is, the evolutionary migration from LPDDR2 to LPDDR3 to LPDDR4, etc., has been achieved mostly by increasing the data rate, as illustrated by Fig. 3.5b, at the cost of power and margin. For mobile applications, this path may not be sustainable, and next generation of DRAM memories is expected to be architected for a wider interface than is used in LPDDR3/LPDDR4—perhaps in some form of a hybrid of the incumbent LPDDR architecture and WideIO2 architecture. Hence, the value propositions described here, based on comparison of WideIO1 and 2 to LPDDR3

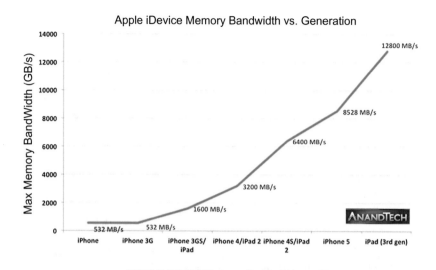

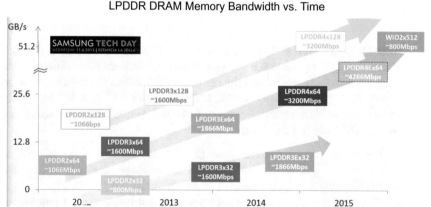

Fig. 3.5 System memory bandwidth trend for mobile applications—plot of system memory bandwidth in apple iDevices (**a**), and LPDDR DRAM component bandwidth versus calendar time (**b**), showing continued growth in demand, and supply, of memory bandwidth

and LPDDR4 DRAM architectures, are generic to any wider architecture and may be applicable to future implementations. Specific value propositions include the following:

i. **Power**: as illustrated in the Fig. 3.6, the power efficiency advantage of WideIO2 versus LPDDR3/4 is significant. Basically, WideIO can deliver a given Bandwidth at much lower power than either LPDDR3 or LPDDR4 architectures. For example, if a BW spec is 25.6 GB/s, system memory power, including memory core, memory I/O, channel and SoC I/O, using WideIO2 is estimated to be ~65% of that with LPDDR4. Of course, the BW demand is dependent on a use case, work load, etc., but, in general, any reasonable

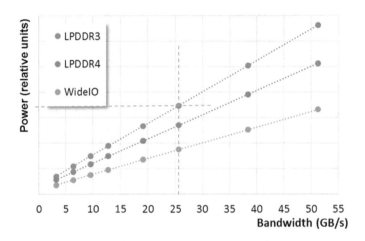

Fig. 3.6 Mobile DRAM power efficiency—a plot of relative DRAM power versus bandwidth for LPDDR3, LPDDR4, and WideIO2 DRAM architectures, with power estimate based on sum of memory core, memory I/O, channel and SoC I/O power components

memory utilization will produce a power differential that is significant. Furthermore, if in the future the Bandwidth requirement for mobile applications goes up to 51.2 GB/s (vs the current $\sim$25.6 GB/s), the absolute memory power difference goes up to $\sim$1 W. 1 W in a phone is a big deal. The specific savings versus PoP implementation of LPDDR3/4—the incumbent solution— are derived from the following:

- Reduction in the channel load (TSV + uBump for Wide I/O versus pad + SoC package + PoP Via + PoP package for LPDDR4 PoP) enabling use of much smaller T2T I/O drivers.
- T2T pins do not face the external world, and hence do not require full HBM ESD protection, enabling even further I/O load reduction.
- With a wide interface, the clock for Wide IO is 1/4 of that required for LPDDR4 (400 MHz vs 1.6 GHz) to deliver the target bandwidth (25.6 GB/s) reducing various losses throughout the system

ii. **Form Factor**: WideIO/TSV implementation also has significant form factor advantages over LPDDR4/PoP implementation—again a value proposition especially attractive for mobile market. The thickness of the TSV package is the lowest currently achievable, with a z-dimension advantage of $\sim$30% over PoP. The TSV package x-y footprint could also be reduced versus PoP package—assuming that the external IO and the associated BGA count allow this—due to elimination of the PoP vias and/or need for compatibility with the standard memory package footprints. This is illustrated in the cartoon in Fig. 3.7 for a 2 GB memory implementation (2 DRAM die w/ 1 GB per die) in PoP for LPDDR4 and with TSV package for WideIO. Clearly, for mobile market this could be a significant value proposition

LPDDR4 PoP on SoC

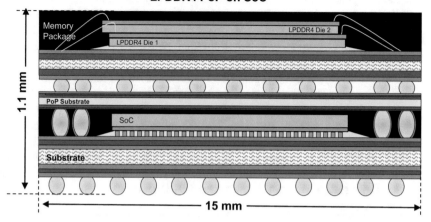

WideIO2 TSV on SoC

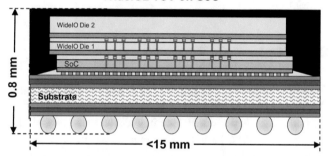

Fig. 3.7 LPDDR/PoP and WideIO/TSV integration—a cartoon illustration of construction of package-on-package (PoP) with LPDDR memory, and TSV with WideIO memory, showing a cross sectional view of 2 GB memory implementation (2 DRAM die w/1 GB per die) on a SoC, and highlighting the typical dimensions

iii. ***Thermal Performance***: normally it is assumed that 3D stacking results in worse thermal performance relative to a 2D planar implementation, since the power density is increased. This is discussed in more detail below in Sect. 3.4.2. However, depending on system implementation and system level thermal management, WideIO/TSV solution in fact could have thermal advantage relative to LPDDR4/PoP implementation. Specifically, since the WideIO solution burns less power for a given bandwidth, it can take longer to reach critical temperature (Chap. 4). In addition, TSV stacking can also take advantage of a top-side heat spreader—a feature that is common in current smartphones. This allows the heat generated in the logic die to flow not only down into the PCB but also up into a system level heat spreader. With PoP this thermal path is inhibited due to the need for overmold in the memory package (to cover the wire bond loops), and the thermal effect of the PoP Side substrate. Thus, with the right system level thermal management solution, 3D SiP with WideIO could have better thermal performance than standard PoP LPDDR implementation.

iv. *Cost*: as shown in the Fig. 3.7, TSS implementation eliminates the need for a memory package, and as such there may be a cost opportunity—all other things being equal. That is, whereas the standard practice is to ship PoP memories in a separate package, WideIO memory must be shipped as KGD die, thereby potentially saving on the cost of a memory package. This is of course balanced versus the cost of the memory die itself; WideIO memory die is larger and requires additional processing (to implement the TSVs for multi-die stacking) than the equivalent LPDDRx die

Ergo, WideIO Memory-on-Logic has a potential Value Proposition, especially suited for mobile applications. Clearly, logic die, and even the entire system, should be designed specifically to leverage this value proposition.

3.4 3D SiP Technology Tradeoffs

The 3D Stacked Die cost-benefit analyzes is at least as complicated as the one for 2.5D Split-Die described in Chap. 2. The degrees of freedom involve incremental, highly inter-dependent, trade-offs that are in addition to the ones described for 2.5D technologies. This section summarizes these trade-offs and the associated key dependencies. Note that the focus is on the trade-offs specific to 3D, in addition to the ones described in Chap. 2 for 2.5D technology.

3.4.1 Architecture Knobs

In order to take full advantage of the value propositions offered by the TSV technology, a chip has to be architected for 3D stacking. That is, it is unlikely that a chip designed for normal "2D" implementation will lend itself to 3D stacking, or be able to leverage the opportunities such as wide T2T interface, low power, etc. Thus, a different set of trade-offs should be made to optimize a product for 3D implementation, versus either 2.5D integration or standard 2D SoC implementation. Principal "knobs" are the following:

i. *System Partitioning*: given the physical implementation constraints, system partitions that truly leverage 3D die stacking technology must be chosen. At a superficial level, it would seem obvious that 3D SiP integration schemes that result in reduced component count by combining multiple components into a single unit (such as Memory-on-Logic, or Sensor-on-Logic, or Analog-on-Logic) can negotiate the conflicting constraints and generate incremental value at the system level (in terms of power, performance, form factor and/or cost) to cover any incremental costs. Similarly, a Split-Die concept implemented in a 3D SiP should have a positive value proposition at a given Si area. However, at a finer level of analysis, there are system level trade-offs that need to be negotiated for any 3D SiP scheme.

Memory integration is an excellent example of complexity of system level analyses required to assess the cost-benefit trade-off of a heterogeneous M-o-L solution. For mobile applications, for example, system memory requirement has ballooned to ∼4 GB for top tier phones, requiring four 1 GB memory die, for memories implemented in 2x, y, z generations of DRAM technology (sweet spot is 8 GB per die). Implementing a unified memory architecture—normally used in mobile operating systems—in a WideIO cube compatible with 3D SiP integration, would then require at least 3 of the 4 DRAM die to carry the incremental TSV area and/or processing. The consequent negative (vs LPDDR wire bond configuration) cost impact may overwhelm the total positive benefits. That is, 4 die WideIO stacks may be prohibitively expensive. One way of offsetting this cost is to use a hybrid, nonunified memory architecture, where a single WideIO memory die is used as a memory dedicated to a subsystem—say GPU, CPU, or Modem—and a lower performance and cost memory (e.g., 3 die of LPDDR3) is used for the rest of the system memory. This would deliver the high bandwidth and power efficiency advantages of WideIO memory where it is needed, and maintain the cost advantages of mature LPDDR memory everywhere else. However, managing the priority and coherency of multiple memory stacks requires an upgrade to the Operating System (aka Non-Uniform-Memory-Access NUMA), and potentially to some application software packages. This then may be a prohibitively disruptive change. Thus, the architecture trade-offs in this example span everything from chip hardware to system software and everything else in-between.

Similarly, a comprehensive assessment of either homogenous or heterogeneous L-o-L is required to define an optimum system partition. Split-Die value proposition (Sect. 2.3.2) could be maximized through use of 3D L-o-L stacking relative to the 2.5D integration—especially if the partition requires more than ∼2000 wires between the die. On the other hand, the impact on the physical attributes of the end component, such as power-performance and thermal characteristics, manufacturability and test, etc., may be negative, and hence must also be assessed. Some examples of different types of trade-offs and constraints are described below.

ii. **Die Size Relationship** (i.e., the relative size of T1 vs T2 die): It is a given that in most cases the value proposition of 3D stacking is maximized when the die (two or more) are approximately of same size, and typically there is a "critical" size below which 3D die stacking does not make economic sense. This is analogous to the critical die size required to amortize the 2.5D Split Die value proposition, described above (Sect. 2.3.2). However, with 3D technology, the *relationship* between the sizes of the 2 die dictates incremental constraints—and hence impacts the feasible architecture-level partitioning. Thus, for example, for Wafer-to-Wafer (W2W) stacking the die sizes must be identical (which, for logic stacks, typically would involve some level of inefficiency in Si utilization, or requires a futuristic design environment capable of implementing very fine grain partitioned design). Similarly, for Die-to-Wafer (D2W) stacking the placement of T2 die must fit within the footprint of T1 die, thereby limiting the freedom of placement of the TSV array on the T1 floorplan. For Die-to-Die (D2D) stacking, typically the smaller die should be on top, thereby dictating that the TSVs be implemented in the technology used to build the larger die, and so on. In addition, as was the case with 2.5D Split-Die, each T2T interconnect precipitates some Si overhead, so that fewer tiers of bigger die may be better than more tiers of smaller die. Some of these trade-offs are further elaborated below.

iii. ***Stacking Order*** (i.e., which die is on top and which on bottom): With 3D, die stacking sequence is also a factor that impacts the architecture. Thus, since the die on bottom (next to the package substrate) must contain the TSVs, the availability and/or economics of TSV process may constrain some die to the top of the stack. For example, for DRAM-on-Logic stacks, the memory typically should be on top, since placement on the bottom would require custom TSV placement and/or processing, thereby destroying the value of a commercial memory. Similarly, for heterogeneous L-o-L stacks, the die implemented in a node that does not include TSV processing must go on top. Note that this sequence of stacking has important implications for the floorplans of both die (TSVs are placement blockage on T1 floorplan and uBumps drive I/O location of T2 die), for the thermal behavior of the stack (high power die on top is preferred for proximity to heat sink), and for PDN integrity (high power die on bottom is preferred for proximity to package/ PCB power rails), etc.

iv. ***Stacking Orientation*** (i.e., which way does the active side of each die face): an additional degree of freedom that must be considered in architecture and design is the stacking orientation: Face-to-Face (F2F) versus Face-to-Back (F2B) versus Back-to-Back (B2B), with "face" referring to the active side of a die. Clearly with more than 2 die in a stack some of these orientations are not possible and F2B is the typically favored orientation. But for 2 die stacks—as illustrated in the Fig. 3.8, with active face highlighted in red—the orientation has an impact on power delivery (PDN delivery directly to the active face of a die is better than PDN delivery through the TSVs from the back of the die), and performance (IO facing the package is better than going through TSVs). Note that the orientation then also has an impact on the total TSV count, and

Fig. 3.8 3D stacking orientation—a cartoon illustration of face to face (F2F), face-to-back (F2B), and back-to-back (B2B) 3D stacking orientations, with "face" referring to the active side of a die (highlighted in red), showing a cross sectional view of concept 3D SiP packages

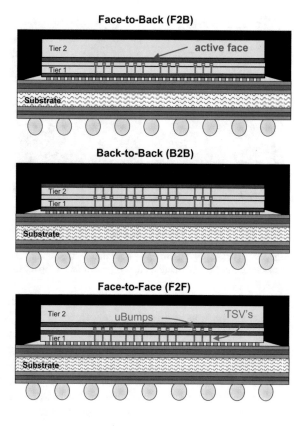

on which wafer may not need to be drilled and thinned, and how much backside processing may be required for which wafer—all of which has a significant impact on cost, etc. There are indeed many tradeoffs...

v. *Stacking Alignment* (i.e., alignment of T2 die relative to T1 die): It is clearly possible to stack the two die so that they are aligned and centered, with T2 die sitting exactly above T1 die—regardless of their respective sizes. It is also possible to offset T2 relative to T1 die, either with T2 confined within the T1 footprint, or even overhang it, with T2 die hanging over the edge of T1, as illustrated in the cartoon in Fig. 3.9. Stacking alignment is a significant variable, directly traded-off versus the choice of the Stacking Process Flow—e.g., offset alignment is incompatible with W2W stacking flow, and overhang alignment cannot work with D2W flow. Furthermore, the yield and cost of the stacking and assembly processes (specifically the underfil process), as well as the mechanical stress magnitude and distribution, are all significantly affected by the die alignment, and relative placement of die edges. Hence, typically the overlap and underlap dimensions are constrained by explicit layout rules, and placement of the T2 die edges can be restricted away from specific T1 features, and vice versa. On the other hand, the design

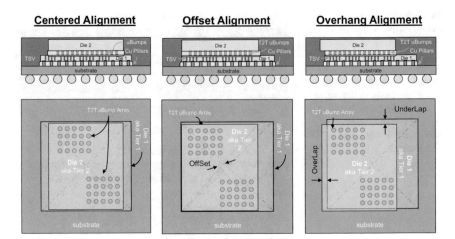

Fig. 3.9 3D stacking alignment—a cartoon illustration of centered, offset, and overhang types of stacking alignments, showing a cross-sectional and a top down view of concept 3D SiP packages

of the die—especially the T1 floorplan—is also impacted by the stacking alignment, as it affects the placement of the TSVs within the T1 floorplan, which in turn impacts T1 utilization efficiency, die size, and cost (see section on TSV-uBump alignment). Thus, this stacking alignment is a key trade-off variable with impact in both process and design domains.

vi. ***I/O Count***: For 3D stacking all external I/O's and the bulk of Power/Ground pins must be accommodated on the bottom die, and all the BGAs must be accommodated on the bottom substrate (in case multiple substrates are used —see below). This requirement may constrain the Tier 1 die size and/or package size, i.e., the minimum die size and package size may be driven by the I/O and BGA count requirements. As such this may constrain the architectural partitions, and dilute the value of L-o-L die splitting and 3D stacking. Conversely, for M-o-L schemes, 3D stacking could produce incremental value by replacing external memory connections with TSV/ uBump connections, and potentially allowing a reduction in both, the die size (by replacing the large LPDDR I/Os with T2T buffers), and package size (by eliminating PoP vias and/or some of the BGAs). Note that, as indicated in Sect. 2.3.3, the number of pins on standard SoCs is showing an increasing trend versus time, i.e., this constraint is likely to be a more important issue with the growing content and complexity of 3D SiPs.

vii. ***Power Delivery***: providing good Power Distribution Network (PDN) to minimize the Vdd droop is a key requirement for high performance ICs. Meeting these requirements is harder with 3D stacks than with standard 2D implementations. In fact, PDN may be a fundamental handicap for 3D, and as such it must be designed in from the very beginning. Clearly, power must be delivered to Tier 2 die (and possibly Tier 1 as well—depending on the stacking orientations) through an array of TSVs. All things being equal, the

Fig. 3.10 Concept dual substrate 3D SiP—a cartoon illustration of a concept 3D SiP using a 2 die B2B stack with dual substrates

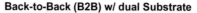

Back-to-Back (B2B) w/ dual Substrate

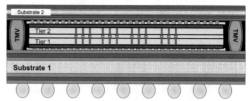

TSV impedance is incremental relative to the standard 2D SoC/FCBGA implementation. The dc performance of the PDN mesh is impacted by the TSV size, fill metal, the number of TSV's in an array, the distance between the current sink, and the TSV array, etc. This, then drives the total TSV count and die stacking (e.g., overlap) rules, which in turn drives the die floorplanning constraints, and the overall Si overhead for 3D stacking.

PDN ac performance—dominated by the inductance of the TSVs, and hence TSV size and wafer thickness—can dictate further stacking sequence and orientation constraints (e.g., high power die on Tier 1 with face down), or, alternatively, may force use of very complex integration schemes—or some mix of these. Thus, for example, to achieve PDN ac performance comparable to the regular 2D Mono-Die, for a L-o-L stack, a B2B configuration with dual substrates can be defined so that each die faces a substrate with good PDN delivery capability (see sketch in Fig. 3.10). But this is complex and expensive and large—potentially destroying the 3D SiP value proposition.

Alternatively, the system must be architected to be resilient to an increased V_{droop}, and/or tolerant to reduced performance. Or, as is most commonly practiced, Tier 2 die in 3D stacks must be restricted to low power/performance content (low di/dt)— for example, WideIO Memory. Thus, PDN is a requirement that imposes fundamental constraints on the architecture of 3D stacks.

viii. ***Performance***: normally 3D stacking is expected to have positive impact on power-performance, since appropriately partitioned system can be implemented with shorter wires, than with either 2D or 2.5D configurations. However, this should be balanced against other constraints—some of which are specific to 3D integration—so that the outcome is not necessarily obvious. For example, with multi-core systems, it is obvious that placing the CPU cores directly above each other in a 3D stack, would result in the shortest wire lengths. However, this 3D floorplan would also align the blocks that typically consume the highest power, with the highest power density, thereby also aligning the thermal hot spots in the stack. This alignment maximizes the local temperature and clearly not only exasperates the thermal management challenges (Sect. 4.3.3), but also makes the PDN design issues harder. On the other hand, interleaving the high power-density layers (such as CPU's) with low power density circuits (e.g., memories serving as local cache), does result in reduced peak local temperatures on the CPU layer, but it also potentially raises the temperature of the more thermally sensitive layer (memories), potentially reducing the net thermal ceiling.

That is, the impact of 3D stacking on system performance must be assessed on case-by-case bases, in multiple domains, and is dependent not only on the global architecture, but also on local physical design details—clearly a key constraint.

ix. *Test*: both, screening out the manufacturing defects and design verification and debug, impose incremental constraints on architecture of 3D stacks; analogous as with partitioning for 2.5D SiP technology. Specifically, in order to test each of the die separately prior to stacking, each die must include adequate functionality and utilities to be testable—as is the case with 2.5D integration. However, the fact that each die must also include the I/Os big enough to drive the load of a standard probe card and to withstand the potential associated ESD events, is a constraint that may inhibit use of some possibly attractive partitions. For example, since with 3D stacking all the external I/Os must be on one of the die (unlike 2.5D), it is tempting to implement the top tier die in a Si technology that does not include the thick gate oxide (necessary for the usual general purpose I/Os facing the outside world, and capable of interfacing to higher voltage rails and supporting standard ESD immunity). Furthermore, whereas the T2T I/O drivers can be designed to be quite small—since they, by definition, drive small loads—the I/O drivers required to support normal test functions may have to be sized up, thereby increasing the Si area overheads. Thus, test requirements eliminate some potentially elegant partitions, and force constraints on both the process technology and the design, at the cost of reduced flexibility in the architecture, more expensive Si wafers, and larger Si die area overheads.

That is, in order to leverage the full potential value proposition of 3D Stacking Technology, the system must be architected for it, i.e., in most cases taking an existing architecture and implementing it in 3D technology would result in a negative value proposition versus mono-die 2D SoC. Furthermore, 3D Technology imposes a number of incremental constraints that must be traded-off and optimized for a given application.

3.4.2   Application Knobs

One of the key differentiators of 3D Technology—versus 2.5D and especially versus 2D—is that it necessarily places two or more die in very close and intimate proximity to each other. The die are, after all, by definition, on top of each other with only a few 10 um's of an insulator separating them. A necessary consequence is that this allows interactions among the die; interactions which can normally be ignored as negligible with the standard 2D SoC, or even 2.5D SiP, implementations. Furthermore, these interactions can be in electrical, thermal and mechanical domains, forcing a requirement for a multi-physics approach to assessing the various trade-offs; something which is oftentimes neglected with the traditional implementation schemes. The physics of the interactions, and the modeling methodologies required to describe them, are described in Chap. 4. This section outlines the basics of possible interactions in 3D SiPs, and focuses on the options for managing them.

i. *Electrical Domain Interactions*: 3D die stacking introduces new features and new interactions, as following:

- *New features*: include the TSV's, the two-sided die with backside metallization (BRDL), and the uBumps on the back of the die, as illustrated in Fig. 3.11a. The characteristics of these features must be modeled to enable design and analyzes. The best modeling approach for describing the nonorthogonal 3D features is some kind of a field solver. For example, uBumps and TSV's are not perfectly cylindrical and as such should be modeled as a series of slices. In addition, the TSV is a

Fig. 3.11 3D technology electrical model—a cartoon illustrating the electrical model associated with the principal new features introduced by the 3D technology (TSV, the two-sided die, backside metallization (BRDL), and the uBumps), and a corresponding equivalent circuit model

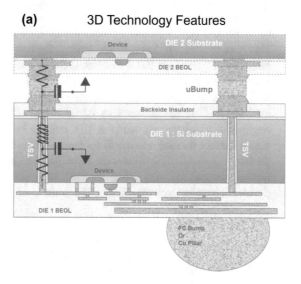

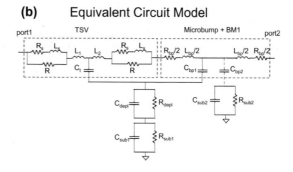

MOS capacitor, and as such has voltage-dependent characteristics dependent on nonuniform (in z direction) oxide thickness and Si doping profile. Thus, puristically, a mesh-based field solver should be used to model the electrical behavior of these features.

However, in practice this is too complex, and the new features must be described as a lumped RCL model—as illustrated in Fig. 3.11b. That is, in practice, a circuit model, as shown (Wu et al. 2012), is defined to describe the electrical characteristics of the new features, that can then be simulated in SPICE. The values of the various RCL elements, of course, must be calibrated versus actual Si measurements.

Note that, strictly speaking, voltage, frequency, and temperature dependencies, as well as process variability should also be comprehended, and some kind of corner models, in addition to the nominal model, have to be defined and calibrated.

- *New Interactions*: include within-die coupling and die-to-die coupling, as illustrated in the cartoon in Fig. 3.12. The intra-die interactions include coupling of traditional planar structures with the new "vertical" structures (e.g., Mx-to-TSV), as well as the interactions among the new features (e.g., TSV-to-TSV). Note that the standard EDA tools used for Si design and model extraction (e.g., SNPS StarRC, MENT xRC, CDN QRC) do not deal with the interactions between vertical and horizontal features (since with standard 2D technologies, the "vertical" features—such as BEOL vias - are tiny and the net impact is negligible). Furthermore, with the TSV being a MOS capacitor, a high frequency signal can inject noise into the Si substrate, which could then create opportunities for long range coupling with some unsuspecting sensitive device. The traditional method for coping with substrate noise is to use guard rings around sensitive devices. However, with the TSVs, the noise is injected deep into the bulk of the

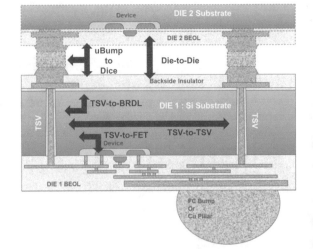

Fig. 3.12 3D technology electrical interactions—a cartoon illustrating the principal electrical coupling modes introduced by the 3D technology features, highlighting the coupling of traditional planar structures with the new "vertical" structures (e.g., Mx-to-TSV), the interactions among the new features (e.g., TSV-to-TSV) and interactions between the stacked die (e.g., ubump-Die)

Si substrate, rather than along the surface, and may thus be coupled into the back of a victim device, so that the guard rings may not be very effective.

A potential solution is to create some kind of a lumped behavioral model to describe these interactions. Note that these models must also be calibrated versus Si measurements—especially for describing the behavior in high frequency ($> \sim 1$ GHz) domain which should be correlated versus suitable S-parameter measurements.

Tier-to-Tier interactions are clearly also possible—especially with Face-to-Face stacks that place active die surfaces against each other. With other stacking configurations, it is normally assumed that the bulk of the Si substrate shields the die from D2D coupling, and this interaction is typically neglected.

That is, simplified, lumped circuit models are required to describe the new features and the electrical interactions possible in TSV-based 3D stacks. These models are not trivial, and a reasonably complex macro, along with the technology-specific and calibrated tech files, need to be included in the SPICE models for the 3D technology. Note that with a distributed supply chain, where the T2 die are built at one foundry, T1 die with the TSV's are built at another foundry, and the reveal processing and uBumping and assembly are performed at an OSAT, the responsibilities for model definition and calibration are also distributed.

ii. **Thermal Domain Interactions**: in general terms, 3D die stacking can be expected to exacerbate thermal issues, since, by definition, the power density per unit volume is increased—relative to a 2D or 2.5D implementation. In most 3D stacking solutions this is true, i.e., 3D stacking does make thermal management more challenging. Furthermore, especially for L-o-L (using die with hot spots) stacks, special care must be taken to ensure that the hot spots on different tiers are not aligned on top of each other, which would amplify any thermal issues. That is, any thermal modeling and mitigation must take in account not only the total power of the 3D stack (along with the lumped thermal resistance of the package), but also the power density distribution across the die, as well as the die-to-die alignment.

However, 3D stacking also opens new opportunities for thermal management. Namely, with heterogeneous 3D stacks that combine a high power layer with a low power layer, such as Memory-on-Logic implementation, stacking can actually offer new opportunities for thermal management.

That is, for heterogeneous 3D stacks, the hot spots (e.g., CPU or GPU blocks) are typically on the logic tier, and hence it is the thermal resistance to this tier that needs to be managed. Consequently, for very high power applications, for systems with forced air or liquid cooling, the logic die needs to be on top of the stack—to minimize the thermal resistance to the heat sink, which is typically attached to the top of the package. Note that for M-o-L stacks, this would then place the memory tier on the bottom, which in turn forces the TSVs drilling in the memory die (typically considered commercially unacceptable).

For lower power applications, such as mobile phones, with no forced air cooling, the PCB and the phone frame serve as the heat spreaders, with the heat ultimately dissipating passively through the phone skins to the environment. Hence putting the logic tier on the bottom of a 3D stack is typically preferred, with memory die on top, and the TSV's in the T1 logic die. However, DRAM memories tend to be especially sensitive to temperature, so that at normal refresh rates DRAMs are typically specified only up to 85 °C (vs ~ 110 °C for logic). Consequently, with a DRAM die on top of logic, thermal decoupling would seem to be a good idea. With 3D TSV-based stacking, the thermal resistance between the tiers is minimal (Si is a good conductor and the insulating layer is very thin at ~ 10 to ~ 20 um), so that thermal decoupling is not practical. Then, the DRAM would seem to be the weak link and dictate the overall thermal performance of the whole stack.

However, if the overmold over the DRAM is thinned down (preferably to zero) and since Si is a good heat conductor the DRAM itself could be used as a heat spreader for the hot spot on the logic die. If a system level heat spreader is attached to the top of the stack, then the 3D stack itself acts to provide a heat path alternative to the usual flow down to the PCB. This is illustrated in the cartoon in Fig. 3.13.

If a system level heat spreader—such as the thermal tapes normally used in modern phones—or some form of a heat sink is provided on the back of the stack, an effective alternative heat path through the stack can be realized, and the temperature of the stack can be brought down to below the DRAM critical temperature. The power of the 3D SiP can then be increased—as illustrated in Fig. 3.14, showing an improvement in the time to critical temperature of an

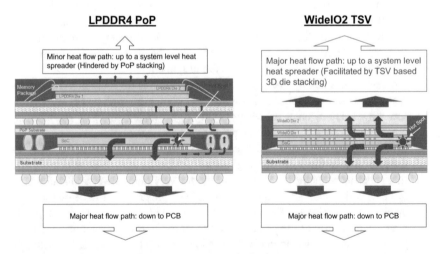

Fig. 3.13 Heat flow in 3D SiP packages—a cartoon illustrating a concept hot spot in a logic die and the flow of heat in a package-on-package (PoP) with LPDDR memory package, and in TSV with WideIO memory package

Fig. 3.14 Thermal mitigation opportunity with 3D SiP—a plot of memory bandwidth versus time to critical temperature in a concept mobile device for PoP/LPDDR and TSV SiP/WideIO implementations, highlighting the potential improvement in thermal performance of the 3D SiP

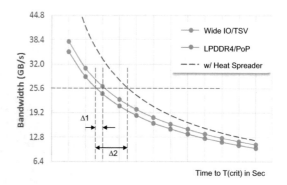

M-o-L stack at 25.6 GB/s by $\sim 10\%$ due to WideIO power advantages (D1), and further $\sim 20\%$ due to the alternative heat path (D2). Note that this heat management solution is not possible with PoP memory, because the thick overmold and memory package substrate used in PoP configurations force a tortuous heat flow path—as illustrated.

Thus, there are tradeoffs that can be implemented at the system and component design level to optimize the thermal performance of a 3D M-o-L stack, resulting in performance beyond the possibilities available with the incumbent PoP solutions.

iii. **Stress**: 3D die stacking technology tends to push the wafer thickness to a minimum, e.g., ~ 50 um for 5 um TSV (or less, as the TSV diameter is scaled). Even at 50 um, the wafer is thin enough to be transparent in the visible light domain and barely has the mechanical strength to hold itself up so that it sags under its own weight (as illustrated in Fig. 3.15). The absence of thick Si substrate, which is normally leveraged to provide mechanical foundation for standard 2D Si devices, exacerbates all stress-strain effects. That is, the use of very thin wafers makes 3D SiPs especially prone to all stress effects and warpage management challenges—described more in Chap. 4.

Stress-Strain phenomena, of course, impact regular planar 2D devices as well —especially when they are thinned down. Products for mobile market currently target ~ 85 to ~ 100 um die thickness; PC/Laptop chips are typically thinned down to few hundred ums. Thus, the TSV wafer thickness of ~ 50 um is slightly beyond the thinness limits currently practiced by the industry—and as such 3D is a good lead indicator of stress-related sensitivities to be expected in this market segment.

Stress-Strain effects are complicated and diverse—both in terms of the sources of stress and the consequences of Strain. The sources of stress can be segregated into "global" and "local" effects, and the consequences of strain can be grouped as "physical integrity" and "electrical performance" effects. Note that this topic, and specifically the design and mitigation methodologies, are addressed in more detail in Chap. 4).

Fig. 3.15 3D TSV 50 um thin wafers—pictures showing that 50 u thin wafer are flimsy and bendable, and are transparent in the visible light domain

- *Sources of Stress*: Stress effects associated with 3D Die Stacking technologies encompass more than the often addressed TSV strain effects (Tengfei et al. 2015; Xu and Karmarkar 2011; Sukharev et al. 2010; Amagai and Suzuki 2010; Deutsch and Lim 2012; Saettlera et al. 2015), i.e., the sources of stress are the following:
  - Global Chip-Package Interaction (CPI), i.e., CTE mismatch between Si die, package substrate, mold, etc., resulting in Die Warpage—exacerbated by the 3D wafer thinness. The usual mitigation avenues include material choices, thermal profile management, relative thicknesses, etc.
  - Local Chip Package Interaction (CPI), i.e., CTE mismatch driven mostly by the volumetric shrink of the underfil, resulting in a local stress peak in the region around the die attach bumps (uBump, Cu pillar/C4 chip bump, and/or

BGA balls). It is exacerbated in 3D technologies due to the thin wafers, and use of small and hard Cu uBumps—acting as stress concentrating pins. It is managed by the usual material and process choices, as well as by placement constraints enforced in design of both Tier 1 and Tier 2 die

- TSV—Si Interaction caused by the CTE mismatch between TSV fill Cu and Si substrate and exacerbated by the relatively large size (at several um this is large for modern Si technologies) of the TSVs embedded in the Si crystal. It is mitigated by suitable management of the in-process thermal hierarchy (to precipitates Cu restructuring and stress relief), as well as by use of the TSV Keep Out Zone design rules (Sect. 4.3.4.)

- Die Edge/Corner Interactions driven mostly by a combination of CTE mismatch and volume change, and exacerbated in 3D by (a) use of thin die and (b) die stacking that creates new and unique edge/corner interactions between the stacked die. Note that some of the die offsetting options, such as die overhang, can create unique stress contours. These effects are mitigated through incremental process constraints and various keep out design and floorplanning rules applied for both Tier 1 and Tier 2 die, as well as the usual material property management.

- *Consequences of Strain* are also multiple and can be seen in both physical and electrical characteristics,

  - Physical: stress-strain effects result in warpage, cracks, delamination, fatigue, creep, etc., that is, material integrity phenomena referred to here as mechanical Chip-Package-Interactions (mCPI). mCPI effects result in catastrophic shorts/opens type of fails that are relatively easily observed. The industry is consequently familiar with managing these, and failure analyzes and fault isolation techniques, and the root physics-of-failure models, have all been developed and are routinely practiced.

  - Electrical: stress-strain effects result in changes in the transistor electrical behavior, potentially causing circuit performance and/or parametric fails— here referred to as electrical Chip-Package/Board-Interactions (eCPI/eCBI). Note that the mCPI phenomena are typically observed at stress levels of the order of low 100s of MPa, but that appreciable carrier mobility shifts can be seen at high 10s of MPa. That is, the eCPI effects may be more sensitive to stress then the traditional mCPI effects. Furthermore, eCPI parametric fails are difficult to identify, characterize and isolate, and mitigation and modeling methodologies are only in nascent phases. eCPI/eCBI is addressed in a separate section below (Sect. 4.3.4).

Thus, in summary, there is a range of complex interactions and phenomena caused by mechanical stress, and mitigation of these effects, exacerbated by the wafer thinness, requires a whole new set of tradeoffs, both in process domain (materials, temperatures), and/or design domain (KOZ rules, placement and floorplanning rules, etc.). A whole new set of knobs and constraints

iv. ***Reliability***: most of the failure mechanisms and reliability mitigation practices developed for 2D apply equally well to the 3D die stacking technologies. The notable exception is "Cu Pumping." This failure mechanism is unique to Cu filled TSV's and is another mechanical stress driven phenomenon. Given the difference in the CTE of Cu ($\sim$16 ppm) versus Si ($\sim$3 ppm), temperature cycling associated with the TSV process, as well as the subsequent BEOL and assembly processing, results in stress buildup around and inside the TSV. The magnitude and polarity (compressive vs tensile) of the stress depends on the specific geometries, material and process conditions. In addition to perturbing the performance of nearby devices at time-zero (eCPI phenomenon mitigated by the KoZ design rules), the vertical component of the TSV stress can also force Cu to bulge out of the TSV, and bend, or even crack, the BEOL interconnect sandwich above it—as illustrated in the micrograph In Fig. 3.16. Most of this stress is precipitated in-line, during the manufacturing process (e.g., BEOL processing), sometimes resulting in time-zero fails. However, with successive thermal cycling, such as could be experienced in field operation, any residual stress can also lead to Cu creeping out of its Si cylinder. Over time, this may disturb the BEOL stack, and ultimately result in metal shorts/opens fails. This is an intrinsic, wear out failure mechanism that is a fundamental reliability concern for Cu filled TSV. Cu Pumping is normally mitigated by managing the thermal cycles through the process, resulting in suitable distribution of Cu grain sizes, and allowing the TSV copper to restructure and stress relive prior to the polishing and BEOL capping (Heryanto et al. 2012; De Wolf et al. 2011; Tu et al. 2016).

Thus, in summary, there are several interaction mechanisms unique to TSV-based 3D stacks, which clearly need to be managed and mitigated, through both, process, and design knobs. Note that none of the interactions are show-stoppers—they are just an incremental set of constraints that must be analyzed and understood, considered, and traded-off versus other constraints, during the design and manufacturing of 3D stacks.

Fig. 3.16 TSV copper pumping—a micrograph of a top portion of a TSV showing Cu bulging out of the TSV and bending the BEOL interconnect sandwich above it

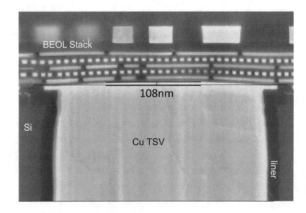

3.4.3   Physical Design Knobs

3D die stacking technology is an opportunity—but the value propositions can be realized only if the die in the stack, and the stack itself, are designed for it. That is, the various partitioning, floorplanning, and routing trade-offs, as well as the technology selections must be optimized for 3D stacking. However, the existing 2D IC design paradigm assumes that a chip is a uniform, isolated, single sided, planar structure that is fully described by a single technology file. This is very different than the reality in a 3D stack that involves multiple technologies and multiple two-sided die, which do interact with each other in physical, electrical, thermal, and mechanical domains. Thus, the design trade-offs involve new methodologies and EDA tools as much as optimizing the various layout constraints. Some specifics:

i. **Design Methodology**: Strictly speaking, design of 3D products requires a new design environment to realize the full 3D SiP value propositions. This is addressed in Chap. 4. However—at this time, a new, 3D design environment is not a realistic proposition, because

   (a)   an alternative 3D design environment does not exist, and
   (b)   the investment in 2D design environment is too large to be abandoned.

Consequently, a design methodology that leverages existing 2D paradigm as much as possible, while allowing for the idiosyncrasies of 3D technologies, is the desired intermediate solution for design of 3D SiP products. That is, in the foreseeable future, the bulk of 3D product design and signoff must be implemented through a 2D Design environment, and the associated 2D EDA tools (thereby preserving the investment and complying with the current realities). This necessitates designing one die at a time. Further optimization can be achieved by developing additional methodologies (and/or tools) which can be used for "off-line" analyzes of a full 3D stack, to ensure product performance and functionality. The next level of enhancement would be achieved by developing a methodology for architecture level analyzes of a 3D stack in order to precipitate the constraints for each individual die, which can then be designed in 2D tools. And, of course, the ideal is to enable a co-design and signoff methodology for a full 3D stack, involving multiple die, multiple technologies, and multiple constraints. That is, there are several levels of evolving the 2D design environment to enable design and signoff of 3D SiPs. For design of Memory-on-Logic type of products, where T2 die and many associated variables are necessarily fixed by the T2 memory design, a basic 2D design methodology, with only a few additions and tweaks, is adequate. The following assumptions and constraints have to apply:

- T2 Memory Die is a commercial product, described by a fixed JEDEC standard => placement of uBumps, drive strengths, frequencies, die sizes, etc., are all fixed and do not need to be optimized
- The content of the T1 Logic SoC is independent of the 3D stacking => no architectural partitioning studies are required

- Partial design verification of the 3D stack is acceptable => no need for integrated DRC, LVS, and Extraction that comprehends both T1 and T2 die in the stack, as well as the structures in between.
- All interactions between the die are ignored in chip level design => manual, off line, assessment of PDN, SI, noise coupling, thermal, and stress effects in the whole stack are done and are used to define incremental floorplan, P&R, and layout rules to be used by the chip level design tools.
- The T2T connections are realized in a straight stack of uBumps and TSVs, i.e., the uBump array defined by the memory die is replicated as a TSV array in the Logic die => the logic die can be considered as single sided die without backside (BRDL) routing
- The TSV and the uBump are all contained within the layout of a T2T buffer => (most of the) chip scale design tools do not need to recognize any of the 3D features at all, but instead do place and route of T2T buffers using the usual LEF, .lib, etc., descriptors of standard cells
- The T2T buffer (including the TSV and the uBump) is defined in a custom layout environment => Custom Layout Tools (e.g., Virtuoso) can be used to design the buffer, including the additional layers and features, with manual characterization to define .lib, LEF, etc.
- The T2T buffer fits within the pitch constraints of the uBump/TSV arrays => IO size and ESD protection must be sized to fit in the x-y constraint of the JEDEC defined uBump array
- Partial test coverage of the T2T buffers during SORT test is ok => no need to probe on uBumps on the back of the logic die, but do need incremental DFT to tie T2T buffer array to chip scan chains

With these constraints, the T1 Logic die for a 3D M-o-L stack can be designed using the usual 2D design environment and tools, with only minor enhancements. Note that these constraints could be extended to other types of 3D stacks as well—for example, stacking analog-on-logic. As long as the content and the design of one of the two die (e.g., the analog chip) in the stack is sufficiently fixed and frozen, so that the assumptions listed above can apply, then the 2D Methodology can be leveraged. The risk, of course, is that this methodology results in sub-optimal stack design, so that the resulting 3D SiP products may not be competitive versus the incumbent solutions.

ii. **TSV-uBump Alignment** (i.e., alignment between the TSV and the uBump): it is clearly possible to layout the 3D stack such that the TSV lies directly underneath the uBump versus offsetting the TSV and connecting them using the backside metallization layer (BRDL), as illustrated in Fig. 3.17 (not to scale). Allowing the offset decouples the placement of uBumps (important for T2 floorplan) and TSV's (important for T1 floorplan). Allowing the offset, however, requires use of backside RDL for routing, rather than just formation of bump pads. Note that offsetting the TSV from uBump also adds the R-C-L of the BRDL interconnect into the electrical path connecting the two die, with

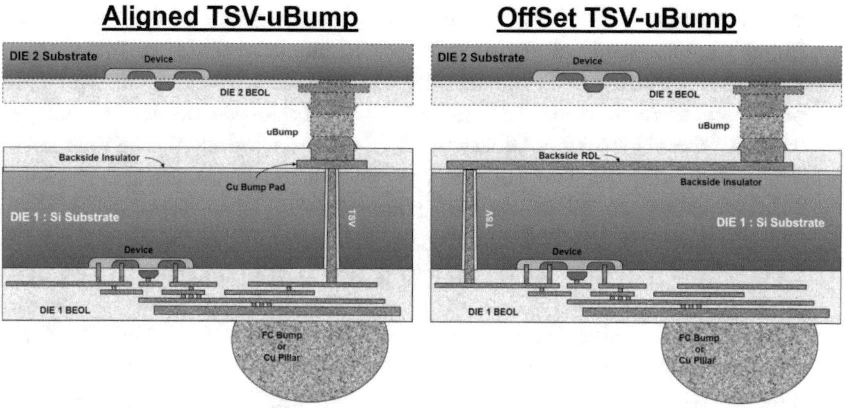

Fig. 3.17 TSV-uBump alignment—a cartoon illustration of Aligned and Offset arrangement between the TSV and the uBump

potential impact on the overall performance, and/or TSV buffer driver size. Furthermore, in order to make T1 floorplan entirely independent of T2 floorplan, which would be enabled by unconstrained offset of uBumps and TSV's, it is likely that multiple layers of backside interconnect are required. This would significantly impact the cost and complexity of backside processing.

uBumps are typically placed in a regular array, rather than being randomly scattering across the die area, in order to make the stacking process more controllable and reproducible. Independent placement of TSVs and connecting them to uBump array, as enabled by the Offset TSV-uBump stacking scheme, would then result in non-uniform, path-dependent, distribution of the parasitic impedance of the BRDL wires. Estimating the path delays then requires automated uBump-BRDL-TSV layout extraction, which in turn necessitates upgrades to suitable EDA tools (Sect. 4.4). On the other hand, use of Aligned uBump-TSV stacking scheme, makes estimating of path delays very simple and uniform, but results in a fixed array of TSVs which needs to be accommodated in T1 floorplan. With large pincounts, and ~ 40 u uBump pitch, the uBump array covers a significant area—of the order of mm's—and mirroring this in a TSV array, results in a significant blockage in T1 floorplan, as discussed below. Hence, relative alignment of TSV versus uBumps is a straight trade-off between flexibility in design, potentially driving the Si die area overheads, versus process complexity, directly driving manufacturing cost. This is clearly a trade-off that needs to be assessed on case-by-case bases.

iii. *Floorplanning*: for a F2B stack, where the T1 die alone includes the TSVs, most of the constraints dictated by the 3D stacking are in fact comprehended just in the floorplanning of the T1 die. Given the design constraints outlined above for M-o-L stacks, the T2T buffer array (including the TSVs), defined to match the uBump array of Tier2 die, must be accommodated in the floorplan of the T1 die. This array may form a large placement blockage in the T1 floorplan.

For example, in a M-o-L stack using WideIO memory as a T2 die, the blockage in the T1 floorplan is of the order of ~ 5.25 mm $\times$ ~ 0.5 mm.

If the IP blocks used in T1 die are designed to accommodate a set of TSVs (i.e., perforated to include the TSVs), then T1 floorplan can be defined independently of the 3D stacking. Otherwise, if IP does not include the TSVs, then the T2T buffer array must be squeezed in-between the IP blocks and accommodated explicitly somewhere in the T1 floorplan.

The placement of the T2T buffer array blockage is a trade-off that must be analyzed. For example, T2 design may prefer to accommodate the T2T buffer array in the center of its floorplan—as is the case with Wide IO memory. However, with similar die sizes and reasonably aligned die stacking, this forces the placement of the T2T buffer array close to T1 die center, which may interfere with the T1 floorplan and force moving the IP blocks toward the periphery. In addition, placing the T2T array in the center of T1 floorplan may adversely affect the routing efficiency, resulting in increased T1 die size. Thus, T1 floorplanning considerations may favor placing the T2T buffer array toward the T1 periphery. This would result in T2 overhang beyond T1 die edge—which must be constrained due to package process limitations. Furthermore, stacking considerations, such as offsetting T2 die away from T1 hot spots, may also favor placing the T2T buffer array toward the periphery of the T1 die. That is, even with the simplified M-o-L type of constraints, the T1 floorplanning requirements can be conflicting, and a trade-off has to be made to identify the optimum without blowing up the chip size.

This is illustrated in the Fig. 3.18, for a specific case study of M-o-L design. "Normal" 2D floorplan needs to be stretched in North–South direction to make a slot between the IP blocks to accommodate the T2T buffer array to the WideIO memory. In addition, T2 die needs to be offset to West, to minimize the routing blockages in T1 floorplan.

iv. **Routing**: The T2T buffer array is likely to be a routing blockage as well as a placement blockage. This is the case with T2T array for WideIO based M-o-L stack, for example. The P/G pins required to supply power to T2 are sprinkled through the array, and need to be integrated with the Logic chip PDN, which is typically distributed through the higher levels of the BEOL stack. Therefore, PDN has to be accommodated in the wiring above the T2T array, thereby consuming some of the routing resources in the area. In addition, the memory signals need to be escaped from the array and connected to the controller on T1. Consequently, the region above the T2T array has low porosity and forms a routing blockage. Thus, putting a routing blockage in the center of Tier 1 is typically awkward and forces the wiring connecting the North with the South side of the chip to be routed around the array, with negative impact on wire length distribution and timing. Same is true for East to West wiring, except that the blockage in that direction is only 0.5 mm versus 5 mm in the North–South direction. Consequently, offsetting the T2 die and moving it off T1 center opens up the routability (toward the West in case illustrated in the Fig. 3.18), and can

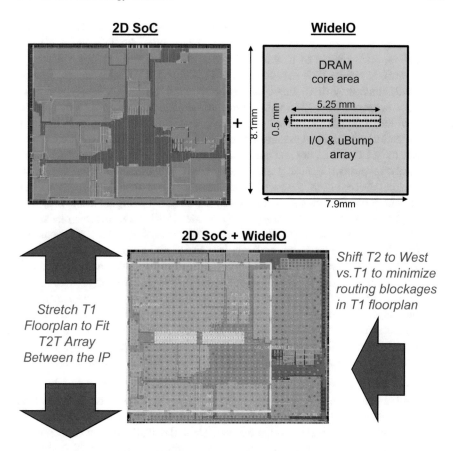

Fig. 3.18 Example 3D floorplan study—a cartoon illustration of an example floorplan of a 2D SoC die and highlighting the alterations required to accommodate 3D TSV stacking of a WideIO memory

have significant impact on chip routability and area utilization, and hence on the overall die size and cost. On the other hand, extending the overhang of T2 over T1 can have significant impact on assembly process complexity and cost and must be within the assembly rules (typically ∼1 mm to 2 mm max offset).

Note that the optimization of the floorplan and routing is very much case specific. The example illustrated in the Fig. 3.18 is specific to WideIO and a given T1 SoC. In general, different SoC content, hard macros, hard macro aspect ratios, routing, and overall die area, etc., can all result in a different optimum trade-off. Obviously, with smaller T1 die it is harder to accommodate large T2T buffer array. Alternative integration strategies—for example, off-setting the uBumps from the TSVs using backside BRDL routing—may be needed. This, however, may require new features in the EDA tools, and the abbreviated methodology outlined above may not be viable.

v. ***T2T Buffer Design***: The 2D Design Methodology used for M-o-L design outlined above assumes that the T2T buffer includes the TSV and ESD protection, and is sized to fit within the uBump footprint (typically ∼40 um × 40 um). However, this may be sub-optimal. The T2T I/O drivers can be quite small, since they drive small loads. Consequently, Signal Integrity and many other concerns associated with normal I/O design are not hard to manage. In fact, the footprint of a 5um TSV itself, and its KoZ, along with the ESD protection structures, dominate the area of a T2T buffer. Since the T2T buffers face only T2, and do not pad out to the outside world, the ESD structure is there purely to protect against possible discharges that can take place during the stacking process itself. As such, these structures serve analogous function as the antenna protection diodes, and need not necessarily be large. Hence, they need not be sized to handle Human Body Model (HBM) discharge—but should manage Charged Device Model (CDM) or Machine Model (MMD) Discharges. The discharge events—if any—during the stacking process are currently unknown, and hence the level of protection required is driven more by fear than fact. Even if there were discharge events during the stacking process, it is unlikely that they would affect T2T buffers in the middle of a dense array, i.e., there is an opportunity to protect the entire array with a few strategically placed structures with pins at the extremes of the array. That is, it is clearly possible to optimize the design of the T2T buffer to be significantly smaller than the uBump footprint, thereby minimizing the placement and routing blockages.

vi. ***Underbump Utilization***: One of the 2D Design Methodology constraints outlined above is that the T2T buffers, including the I/O driver, the TSV, the DFT infrastructure and the ESD protection, are assumed to be sized to consume the entire area in the shadow of the uBump. Thus, the availability of Si area in the shadow of the uBump array, for anything other than the T2T buffers, is assumed to be zero. As per above, it is clearly possible to optimize the size of the T2T buffer, and increase the placement porosity of the T2T array, to accommodate incremental logic within the region. This then reduces the Si area overhead associated with the 3D stacking. That is, with 3D it is possible to improve the use of the area under the uBump (underbump utilization), by optimizing the T2T buffer design. Note that this is difficult to accomplish with the 2.5D integration, because the D2D buffer array is necessarily placed along the die edge—where increased porosity or logic area has low value (Sect. 2.4.2). Hence 3D tends to incur lower Si area overhead than 2.5D because it is possible to get better underbump utilization. This becomes significant if the pin count is high—in which case factors such as TSV size, uBump pitch, etc., may all have a different trade-off point. Analyses indicates that with the 40 um uBump pitch, and 5/50 TSV, 3D is more efficient than 2.5D when the pin count goes >∼2000.

3.4.4   TSV Technology Knobs

The basic process steps for formation of the TSV and integration into the Si and Packaging process flows are discussed in Sect. 3.2. This section focuses on the incremental trade-offs in the process integration domain. These knobs are mutually dependent and are constrained by a few basic parameters summarized here:

i. *Aspect Ratio* (i.e., ratio of TSV depth to its diameter): is a critical and basic characteristic of the TSV formation process. Clearly, maintaining the footprint of the TSV at a minimum is highly desirable as this minimizes the impact of TSV insertion on T1 floorplan, and therefore minimizes the Si area overhead associated with 3D technology. This favors a high aspect ratio TSV. On the other hand, high aspect ratio adversely affects the difficulty of each of the critical steps (etch, liner, fill, reveal) in TSV formation process—and hence the integrity and cost of the 3D SiP. Aspect ratio of the order of 10:1 is currently taken as the sweet spot of TSV formation process—especially for the TSV-middle flavor.

ii. *Wafer Thickness*: Given that the TSV aspect ratio is, in practice, limited to about $\sim 10{:}1$ or less, wafer thickness is limited to $\sim 10x$ the TSV diameter. The desire for the minimum TSV footprint then favors use of very thin wafers. However, thin wafers are very fragile and awkward to handle through the packaging and assembly processes, or require additional bonding/debonding steps to attach them to temporary carriers, or need to be thinned after permanent attach to a carrier. These considerations therefore favor use of thick wafers—driving the TSV area overhead up. Hence the selection of the wafer thickness and TSV diameter—which in turn dictate much of the rest of the process technology—is critical and is related to the target application and the 3D ecosystem. For example, TSV via-middle technology seems to have asymptoted toward via diameter of the order of ~ 5 u, with wafer thickness of the order of 50 um (referred to as 5/50 TSV), and TSV's used in Si interposers tend to favor 10 u vias with 100 u thick wafers (10/100 TSV)

iii. *TSV Keep Out Zone* (KoZ): TSV is a "placement blockage," i.e., CMOS devices cannot be placed in the footprint of a TSV. In addition to the footprint itself, the blockage area includes a "Keep Out Zone" (KoZ) around the TSV—usually approximately equal to the TSV diameter. The keep out zone is necessary to mitigate the effect of mechanical stress associated with TSVs—caused by the mismatch in the TCE between Cu (16 ppm) and Si (3 ppm) on CMOS device performance (Sect. 4.3). With technologies that use W fill, this is much less of an issue since its TCE (4.3 ppm) is closely matched to that of Si. The net effect of a KoZ that is approximately equal to the TSV diameter is that the placement blockage in the active circuit is greatly enlarged—going from ~ 20 um^2 for 5 u via, to ~ 175 um^2. However, this exclusion zone may be used for interconnect, diodes, resistors, ESD protection structures, and so on; that is, it is not necessarily just white space. Nevertheless, technology parameters that affect the KOZ—including fill material, process thermal history, capping materials, device orientation and polarity, device channel length, etc., all therefore have a significant impact on the overall TSV Si area overheads

iii. ***uBump size and technology***: the bumping and die attach technology is obviously as essential to 3D stacking as the TSV technology, i.e., for most applications there is no 2.5D or 3D integration without fine grain die attach technology. Currently the practical limit in bump pitch is of the order of ~ 30–40 um pitch, with ~ 20 u to 30 um bump diameter. In addition to the bump itself, a solder layer, with all the barrier films required to manage the Inter-Metallic Compounds (IMC) integrity and joint reliability, is necessary to form the bond. The attach process is typically either thermocompression bonding (TCB) or mass reflow. TCB is typically required to manage the warpage and planarity issues. Mass reflow process, however, has more desirable cost structure.

Lab studies (De Wolf et al. 2011; Tu et al. 2016; De Preter et al. 2015) indicate that there is no intrinsic reason why uBump technology could not be scaled down toward ~ 10 u pitch. However, for implementation in manufacturing the tradeoffs are complicated and involve test and probing technology, ESD protection requirements, drive strength, and performance requirements, backside RDL and uBump placement decisions, etc. The pitch associated with a T2T I/O, including the 5 um TSV and CDM type of ESD protection, is of the order of 25–30 um, so that there may be no benefits in terms of 3D stacking Si area overhead, in pushing the uBump pitch much below the current level. On the other hand, of course, a fine grain partitioned design, requiring millions of T2T connections, cannot be implemented without a much finer interconnect pitch—believed to be beyond the capabilities of bumping technology, and more in line with the characteristics of a hybrid bonding technology.

iv. ***Backside Processing***: Another uniqueness associated with 3D technology is that it necessitates Si wafers to be double-sided slices, rather than a simple planar structure that the conventional 2D technology is used to. That is, a metallization is required on the backside of the wafer—at least to create a pad for the uBump. With the Aligned TSV-uBump stacking scheme (Sect. 3.4.3) the minimum backside processing requirement is for patterned dielectric (to isolate the pads from Si substrate) and a metal layer (to define the uBump pads), i.e., 2 masking layers. Note that this processing necessarily must be done after thinning of the wafer required to reveal the TSV. It is clearly possible to extend the use of the single backside metallization layer for redistribution interconnect —enabling some backside routing required to allow some limited offsetting of TSV from uBumps. This would add some minor, incremental cost. Multiple layers of BRDL routing, required to enable full decoupling of uBump placement from TSV placement, would however have a major impact on process cost. Thus, there is a trade-off; backside RDL (BRDL) routing may enable reducing the overall Si overhead area on either Tier 1 or Tier 2, or both, but at the price of increased process cost.

v. ***Underfil and Mold***: underfilling 3D stacked die—either capillary underfil (CUF) or mold underfil (MUF)—is also a special challenge, since with the 40 um uBump pitch, the gap between the die to be underfilled is ~ 20 um. An underfil process flow that produces no voiding, with acceptable bleed-out beyond the die edges, is challenging and is a part of the critical tradeoffs that define the stacking flow. Furthermore, underfil material properties and process temperature has a big impact on the residual stresses in the 3D package—which in turn impact warpage and planarity (mCPI), and performance and reliability (eCPI). In addition, the underfil material and process needs to be harmonized with the molding process, as the stresses at the underfil-mold boundary may cause issues—again especially in the case where T2 overlaps T1. Note that due to the thermal considerations the overmold may need to be reduced to zero—allowing the back of the die stack to be exposed—which may exacerbate the overall warpage control challenges.

In summary, optimum use of 3D die stacking technology, even when extremely constrained, such as for Memory-on-Logic stacking, involves many degrees of freedom, and many new interactions—all of which call for complex and multi-domain tradeoff analyses. These tradeoffs necessarily involve diverse considerations, ranging from the capabilities of EDA tools, the nature of the design, the application space, etc., all beyond the realm of just the Si and Packaging technologies. It is complicated.

3.5   3D SiP Technology Solutions

Clearly, 3D integration technology—specifically the TSV based technology options discussed above—offers many degrees of freedom and there are multiple ways that the technology can be leveraged to create an integrated 3D SiP. The sweet spot that maximizes the possible value propositions is application dependent.

In general, the tradeoffs are inter-dependent, and 3D technology seems to be evolving into several types of ecosystems, where one basic selection, in one given domain, determines the optimum choices in different domains. For example, to illustrate the point, from the manufacturing process point of view, W2W stacking is attractive—it is relatively low cost, and simplifies the integration flow (e.g., the many of the carrier bond/debond steps). However, W2W stacking also mandates that the 2 die are of equal size. Consequently, this stacking scheme is attractive for homogenous M-o-M stacking. On the other hand, this requirement, along with the constraints in the design environment, makes this stacking scheme unattractive for heterogenous integration where the die is almost necessarily not the same size. Hence heterogenous integration favors D2D or D2W stacking. Table 3.2 summarizes some of the generic groupings into these 3D ecosystems.

Table 3.2 3D integration technology ecosystems—a summary of the combinations of key 3D technology attributes and constraints for various SiP integration schemes

	Homogeneous		Heterogeneous	
Integration scheme	M-o-M	L-o-L	M-o-L	X-o-L
Stacking flow	W2W	D2D / D2W	D2D	D2D
Die sizes	T2 = T1	T2 < T1	T2 =< T1	T2 =< T1
Stacking alignment	Centered	Centered or offset	Overhang	Offset or overhang
Backside metallization	Pad only	2LM routing	Pad only	2LM routing
Design methodology	N/A	Full 3D	mostly 2D	Extended 2D

Most of the industry experience to date with 3D stacking is with the homogenous M-on-M stacking, and heterogenous M-on-L stacking. Whereas there has been a lot of work in academia (Thorolfsson et al. 2010; Chan et al. 2016; Kim et al. 2012; Jung et al. 2015; Franzon et al. 2013; Neela and Draper 2015; Loh et al. 2007) focused on homogenous L-o-L and heterogenous X-o-L ("X" being an undefined layer that could be Logic, or Analog, or Sensor) integration, there has not been a lot of industrial experience with 3D SiP L-o-L or X-o-L products in volume manufacturing. This section describes the current status of the major 3D integration schemes.

3.5.1 3D Technology Solutions for Heterogeneous M-o-L Integration

A generic concept process integration flow for building an M-o-L type of a 3D SiP is illustrated in Fig. 3.19. The illustrated flow is a two-step process developed for WideIO implementation, with a Die-to-Substrate (D2S) attach, followed by a Die-to-Stack (D2St) attach.

Also, as discussed above (Sect. 3.2), the Post Wafer Fab Flow, including the wafer thinning, TSV revealing and backside processing, is shown to be performed at either the foundry or the OSAT. This is analogous to the MEOL processing for 2.5D integration. The flow also illustrates both, use of a single memory die or a memory cube, within the memory vendor section.

There are of course, possible variations of this basic concept integration flow, with opportunities for, for example, a Die-to-Wafer (D2W) followed by Stack-to-Substrate attach. The best integration solution depends on a specific stack, but clearly, the flow involves two or more attach steps (not counting the stacking of the memory die into a cube), which need to be sequenced to optimize the cost and mitigate yield, warpage, test, etc., concerns.

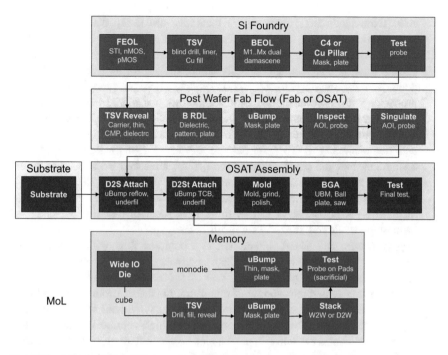

Fig. 3.19 Generic 3D TSV-based Memory-on-Logic assembly flow—a generic high level assembly flow for a 3D SiP package, including the major components and highlighting the essential manufacturing and assembly steps for a WideIO Memory-on-Logic stack

The integration flow shown is derived from the experience with development of a WideIO M-o-L solution. The general conclusions and expectations based on this experience are summarized below:

i. *TSV Formation*: the basic TSV process, including TSV etch, liner and fill steps are by now relatively mature, and no intrinsic issues seem to have been flagged. Use of TSVs for M-o-M stacking and for 2.5D integration—processes that are qualified and in production—have further proven out the core TSV processes. It is believed that the inspection and metrology techniques still need to be enhanced in order to be effective screens for manufacturing process control. Nevertheless, the TSV formation process is yielding at levels compatible with volume production. The opportunities for future development are expected to be mostly in driving the process cost down, and scaling the TSV dimensions—preferably without reducing wafer thickness.

ii. *Stacking Process*: similarly, the uBump-based attach stacking and assembly process has been developed, is qualified, and in production at a number of memory vendors and OSATs. It is believed that currently the underfill process steps, using liquid (CUF or MUF), paste (NonConducting Paste—NCP) or solid (NonConducting Film—NCF) materials, are likely the steps with the lowest process margin, and hence the opportunities for future development.

iii. ***CMOS Integration Flow***: the integration of the 3D modules with the basic CMOS process has been successfully demonstrated; for logic down to 20 nm node, and for memory to 2z and 1x nodes. However, integration of TSV processes with more advanced logic technology nodes, at 14 nm FinFET and below, has not been demonstrated. In fact, it is believed that in the absence of a specific product opportunity, at this time, no foundry is working on developing and demonstrating a TSV module in advanced CMOS logic nodes. The equipment, and the basic process modules for advanced TSV processing are available, and have been demonstrated in a lab (e.g., IMEC 2016), but integration with the advanced manufacturing process flows—other than DRAM memories—is not pursued at this time. It is possible that the Backside via technology deployed at OSATs will fill in the gap

iv. ***Supply Chain***: 3D M-o-L integration is a complex flow with activities distributed across multiple Supply Chain partners (foundry, memory vendor, substrate supplier, OSAT). In fact, one of the reasons for WideIO memory missing the commercial window of opportunity is the distributed nature of the supply chain—without a clear "owner" of the risks involved with deploying a new and disruptive solution. In the absence of a high-volume WideIO product, yield, and manufacturability issues cannot be quantified, and 3D M-o-L integration is hence still perceived as a relatively high risk technology. However, with M-o-M products, such as HBM and HMC, ramping into production, these concerns should be reduced in the future. Note that with the attach of the DRAM cube on a memory management die, both memory architectures also include an M-o-L stacking step.

v. ***Design***: As discussed above (Sect. 3.4.3), a Design Methodology, with a relatively minor enhancement to the standard 2D Tools and Flows is adequate for design of M-o-L type of products. The biggest risks associated with this flow are the implementation of automated checks for compliance with the various KoZ rules (including the SI and eCPI rules and guidelines for placement around TSV and uBump features, and around underlapping and overlapping die-on-die edges). Note that some of these checks need to be implemented at floorplanning steps, as well as the final signoff at GDS2 level.

vi. ***Product Value Proposition***: Ultimately, 3D technology will be used in a SiP product if and when it makes commercial sense, relative to an incumbent solution. WideIO M-o-L SiP concept targeted the mobile market, where the incumbent solution is PoP implementation with LPDDRx DRAM memory. An architectural comparison that assessed multiple possible physical implementation schemes, and evaluated each for multiple figures of merits concluded that each architecture (LPDDR PoP vs WideIO 3D SiP) had different strengths (Radojcic et al. 2014, 2015), as summarized in the Table 3.3. Thus, Wide IO

Table 3.3 LPDDR PoP versus WideIO TSV SiP implementation—a summary of a direct comparison of a LPDDR PoP versus WideIO TSV SiP integration schemes, identifying the superior solution for use in high end mobile devices for several selected metrics

Metric	Best in class	Comment
Cost	PoP	Component cost (AUC) and system cost (BOM)
Bandwidth	-par-	Based on memory manufactures spec'd BW ratings
Density	-par-	Sum of DRAM bits w/standard 0.5 and 1 GB die
Thickness	3D SiP	Worst case component height estimates
Thermal	3D SiP	Thermally limited performance, i.e., time to Tcrit
Power	3D SiP	Sum of static power estimates for a blend of use cases
Biz model	PoP	Ease of manufacturing and monetizing in supply chain

3D SiP has superior thermal performance, lower power, and better form factor, but LPDDR PoP solution is lower cost and is lower risk to implement in a distributed supply chain. Clearly, the industry favors the cost and business model advantages over other value propositions.

Note that when such a comparative assessment is performed at the system level —based on a given smartphone implementation—the effect of the differences between the two architectures are diluted, resulting in advantages/disadvantages of only a few %. That is, whereas a direct, component level, WideIO to LPDDR comparison can highlight differences of 50% in power, or 30% in thickness, (Samsung 2012; Kim et al. 2011) etc., when the entire system is taken in account, the difference is down to few %.

The cost-difference is in fact dominated by the cost of the memory die itself— especially if WideIO is implemented in a cube, thereby incurring the additional cost of TSV processing and uBump stacking. However, the price of commercial memories is dominated by market dynamics, and technology maturity, such that, if and when WideIO is implemented in manufacturing, the cost-difference versus LPDDR is expected to decrease over time.

3.5.2 3D Technology Solutions for Homogeneous and/or Heterogenous L-o-L Integration

A commercially successful 3D SiP L-o-L product could be possible as the intrinsic value proposition and the basic technology modules are there. However, the design environment required to fully realize the value proposition and to optimize the architecture and design of the two (or more) die, does not exist at this time, as discussed in Chap. 4. Hence, L-o-L is, at this time, more or less an academic proposition. The value propositions and the necessary trade-offs have been described above, and L-o-L implementation remains as an opportunity to be realized in the future.

References

Amagai M, Suzuki Y (2010) TSV stress testing and modeling. In: IEEE 60th electronic components and technology conference (ECTC)

Amkor, Flip Stack CSP. http://www.amkor.com/go/packaging/all-packages/flipstackand174-csp/flipstackand174-csp. Accessed 25 Nov 2016

ASE, SiP|Hybrid (W/B + Flip Chip). http://www.aseglobal.com/en/Products/4-1-6-7.asp. Accessed 25 Nov 2016

Braun T (2016) Foldable fan-out wafer level package electronic components and technology conference (ECTC), 2016 IEEE 66th

Chan WT et al (2016) Revisiting 3DIC benefit with multiple tiers. In: 18th system level interconnect prediction workshop (SLIP)

De Preter I et al (2015) Surface planarization of Cu and CuNiSn micro-bumps embedded in polymer for below 20 μm pitch 3DIC applications. In: AMC 2015—advanced metallization conference

Deutsch S, Lim SK (2012) TSV stress-aware ATPG for 3D stacked ICs. In: IEEE 21st test symposium (ATS)

De Wolf I et al (2011) Cu pumping in TSVs: effect of pre-CMP thermal budget. Microelectron Reliab 51(9–11)

Farcy A et al (2013) ST strategy on 3D integration. LSI January 23, 2013. http://www.semi.org/eu/sites/semi.org/files/docs/ST%203D%20strategy%20Semi%20Janv%202013%20diffx.pdf. Accessed 25 Nov 2016

Franzon PD et al (2013) Exploring early design tradeoffs in 3DIC. In: 2013 IEEE international symposium on circuits and systems (ISCAS)

Fujitsu, Stacked MCP. http://www.fujitsu.com/downloads/MICRO/fma/pdf/stackedmcp.pdf. Accessed 25 Nov 2016

Gu SQ et al (2012) 3D integration of wide IO memory cube stacking to 28 nm logic chip with high density TSV through a fabless supplier chain. In: International symposium on microelectronics: FALL 2012

Heryanto A et al (2012) Effect of copper TSV annealing on via protrusion for TSV wafer fabrication. J Electron Mater 41(9)

IMEC, 3D system integration. http://www.imec-nl.nl/nl_en/research/cmos-scaling/3d-system-integration.html. Accessed 25 Nov 2016

IME, Interconnects and packaging (IPP). https://www.a-star.edu.sg/ime/RESEARCH/INTERCONNECT-AND-PACKAGING-IPP.aspx. Accessed 25 Nov 2016

JEDEC (2011) Wide I/O single data rate, JESD229. https://www.jedec.org/. Accessed 25 Nov 2016

JEDEC (2013) High bandwidth memory (HBM) DRAM, JESD235. https://www.jedec.org/. Accessed 25 Nov 2016

JEDEC (2014) Wide I/O 2 (WideIO2), JESD229-2. https://www.jedec.org/. Accessed 25 Nov 2016

Jung M et al (2015) Fine-grained 3-D IC partitioning study with a multicore processor. IEEE Trans Compon Packag Manuf Technol 5(10)

Kim DH et al (2012) Block-level 3D IC design with through-silicon-via planning. In: 17th Asia and South Pacific design automation conference (ASPDAC)

Kim DW et al (2013) Development of 3D through silicon stack (TSS) assembly for wide IO memory to logic devices integration. In: IEEE 63rd electronic components and technology conference (ECTC)

Kim J-S et al (2011) A 1.2 V 12.8 GB/s 2 Gb mobile wide-I/O DRAM with 4 × 128 I/Os using TSV-based stacking. In: IEEE 2011 international solid state circuits conference (ISSCC)

LETI, CoolCube ™: a true 3DVLSI alternative to scaling. http://www-leti.cea.fr/en/How-to-collaborate/Focus-on-Technologies. Accessed 25 Nov 2016

Loh GH, Xie Y, Black B (2007) Processor design in 3D die-stacking technologies. Micro IEEE 27 (3)

Merritt R (2016) TSMC Expands its 3D Menu, EETimes, 9/22/2016. http://www.eetimes.com/document.asp?doc_id=1330503. Accessed 25 Nov 2016

Micron, Hybrid memory cube. https://www.micron.com/products/hybrid-memory-cube. Accessed 25 Nov 2016

MIT Lincoln Lab, 3D integration of CMOS and other integrated circuit technologies. https://www.ll.mit.edu/mission/electronics/qiin/cmos-technology/3d-integration-of-CMOS.html. Accessed 25 Nov 2016

Monolithic 3D, The most effective path for future IC scaling. http://www.monolithic3d.com/. Accessed 25 Nov 2016

Neela G, Draper J (2015) Congestion-aware optimal techniques for assigning inter-tier signals to 3D-vias in a 3DIC. In: 3D systems integration conference (3DIC)

PTI, 3-dimensional integrated circuit (3DIC) packaging. http://www.pti.com.tw/ptiweb/D0011.aspx?p=D&c=D11. Accessed 25 Nov 2016

Radojcic R et al (2014) 2.5D & 3D integration: where we have been, where are we now, where we need to go. In: 4th annual global interposer technology (GIT) workshop, Atlanta, Georgia

Radojcic R et al (2015) 2.5D & 3D integration: where we have been, where are we now, where we need to go. In: iMAPS 11th international conference and exhibition on device packaging fountain hills, Arizona USA March 17–19, 2015

Saettlera M et al (2015) μ-Raman spectroscopy and FE-modeling for TSV-stress-characterization. Microelectron Eng 137

Samsung (2012) 3D TSV technology and wide IO memory solutions. http://www.samsung.com/us/business/oem-solutions/pdfs/Web_DAC2012_TSV_demo-ah.pdf. Accessed 25 Nov 2016

Samsung (2013) Samsung widcon technology. https://www.bing.com/videos/search?q=exynos%2ctsv%2c+wide+io%2c&qpvt=exynos%2ctsv%2c+wide+io%2c&view=detail&mid=6D298AB51163564AFA1A6D298AB51163564AFA1A&FORM=VRDGAR. Accessed 25 Nov 2016

Samsung, Achieving unprecedented capacity through TSV technology. http://www.samsung.com/semiconductor/products/dram/server-dram/. Accessed 25 Nov 2016

SK Hynix, HBM. https://www.skhynix.com/eng/product/dramHBM.jsp. Accessed 25 Nov 2016

Smith M, Stern E (1964) Methods of making thru-connections in semiconductor wafers. USPTO US 3343256, filed 12-28-1964, granted 9-26-1967

Sony, Sony develops next-generation back-illuminated CMOS image sensor. http://www.sony.net/SonyInfo/News/Press/201201/12-009E/index.html. Accessed 25 Nov 2016

StatsChipPac eWLB (FOWLP Technology). http://www.statschippac.com/services/packaging services/waferlevelproducts/ewlb.aspx. Accessed 25 Nov 2016

Sukharev V et al (2010) 3D IC TSV-based technology: stress assessment for chip performance. AIP Conf, Proc 1300

Tengfei J et al (2015) Through-silicon via stress characteristics and reliability impact on 3D integrated circuits. MRS Bull 40

Tezzaron, DiRAM4™ 3D memory. http://tezzaron.com/. Accessed 25 Nov 2016

Thorolfsson T et al (2010) Logic-on-logic 3D integration and placement. In: IEEE international conference on 3D system integration, 3DIC

Tu KN et al. Electronic thin film lab at UCLA. http://www.seas.ucla.edu/eThinFilm/index.html. Accessed 25 Nov 2016

West J et al (2012) Practical implications of via-middle Cu TSV-induced stress in a 28 nm CMOS technology for wide-IO logic-memory interconnect. In: 2012 symposium on VLSI technology (VLSIT)

Wu X et al (2012) Electrical characterization for intertier connections and timing analysis for 3-D ICs. In: IEEE transactions on very large scale integration (VLSI) systems, vol 20, issue 1

Xu X, Karmarkar A (2011) 3D TCAD modeling for stress management in through silicon via (TSV) stacks. In: AIP Conference Proceedings 1378

Chapter 4
More-than-Moore Design Eco-System

4.1 Overview

With the predictable paradigm of More-Moore scaling, technology and architecture development proceeded pretty much in parallel. Up to the very advanced FinFET generation of technologies, the technology community did not require a lot of detailed inputs from the system architects to invest the effort to develop the next CMOS node. And vice versa—the architects did not need to have a lot of detailed inputs from the technologists to develop suitable architectures that leverage increasing device counts. MM Process and design obviously do intersect and interact, especially for layout of the libraries and other lower level IP—and the actual product implementation is of course very technology-specific, based on process-optimized PDKs, libraries and so forth. But, products could be architected at a high level before the technology was qualified—because the basic characteristics of the technology were predictable. And technologies could be developed based on something like ITRS Roadmap, without product-specific specs.

With the More-than-Moore paradigm, the situation is a bit different. In the absence of a tight definition of a MtM technology roadmap, and in the current environment of proliferating technology options, architecture and design must be coordinated with technology development. In the absence of a roadmap and/or historical experience, the tradeoffs in architecture, design, and MtM technology space, as well as the product schedule, AUC, ROI, etc., predictions must all be made on case by case bases (as discussed in Chaps. 2 and 3). This has been a topic of discussion in the industry for some time and many approaches have been proposed (Xie et al. 2011; Papanikolaou et al. 2011; Wu et al. 2011; Topaloglu 2015; Elfadel et al. 2016).

© Springer International Publishing AG 2017
R. Radojcic, *More-than-Moore 2.5D and 3D SiP Integration*,
DOI 10.1007/978-3-319-52548-8_4

The situation is analogous to the practices used by the Si Industry several decades ago—before the cost of fabs ballooned and ITRS became entrenched—and still used by some segments of the industry (e.g., analog or MEMS products?). As long as the design process coordination is a case-by-case exercise, an entire product sourcing eco-system is required for a MtM technology option to intersect a commercial product implementation—in order to mitigate the risks associated with the change from the incumbent paradigm. This eco-system must *at least* include the following elements

(a) a methodology for coordinating design and technology development activities, and
(b) a design environment that can work with the features that an MtM technology offers,

This section discusses the current status of 2.5D and 3D SiP product realization eco-system, including design and process co-development methodologies and EDA tools, and proposes some options and opportunities. It is believed that this eco-system is an existential issue for 2.5D and 3D technologies, i.e., it will be difficult to realize competitive commercial 2.5D and 3D L-o-L SiP products without such an eco-system—at least within the foreseeable future.

4.2   More-than-Moore Architectural Opportunities

Adoption of 2.5D or 3D technologies to realize SiP products is afflicted by the chicken-and-egg conundrum. Product Managers are reluctant to adopt a new and disruptive technology in the absence of demonstration of its cost, yield, and risk attributes. On the other hand, it is difficult to demonstrate the cost, yield, and risk attributes in the absence of a product pushing volume through the manufacturing line. Similarly, product managers are reluctant to adopt a new technology paradigm in the absence of a suitable design eco-system, but the EDA vendors hesitate to invest the effort to develop a new design eco-system in the absence of committed products, etc. Consequently, an iterative methodology of co-optimizing technology options and architectural choices—along the lines described in Chaps. 2 and 3—while developing both, has to be used.

Note that this chicken and the egg conundrum is not necessarily applicable to a situation when the incumbent technology encounters a brick wall, and target product specs just cannot be met within the existing paradigm. Examples of such brick wall would be a situation where mono-die implementation exceeds mask reticle size, thereby forcing a split die paradigm; or an existing memory paradigm not meeting the required bandwidth demand forcing the use of something like HBM or

HMC memories, etc. That is, in a case when the incumbent paradigm runs out of gas, necessity drives adoption of a new paradigm, and may well end up using 2.5D or 3D integration technologies.

In the absence of a compelling situation like that, the chicken and the egg conundrum holds and a 'concept architecture' is required to estimate the technology value proposition and to explore its various tradeoffs. In that case, the conclusion of an assessment may be dependent on the concept architecture itself, thereby making pathfinding an iterative loop. For the purposes of this book and More-than-Moore technologies 'pathfinding' is defined as: "a procedure to find an optimum solution for a previously unexplored technology paradigm". Some of the dependencies of pathfinding on the concept architectures and concept technologies are described in this section.

4.2.1 Architecture Concepts

Advent of a disruptive technology—such as the 2.5D and 3D integration technologies—may be an opportunity to explore disruptive architectural concepts as well. Some of the possible concepts are summarized here—more to illustrate the idea of leveraging an entirely different approach, than to propose an economically and/or technically viable solution.

i. *Split memory architecture*: As briefly described in Chap. 2, functional partitioning of a mono-die SoC into a 2.5D multi-die SiP may be dominated by constraints associated with access to a unified system memory. Thus, the best partition may be to allocate all the blocks that are sensitive to memory latency on the same die as the DRAM memory controller—potentially ending up with CPU and GPU on different die.

An alternative partitioning solution, where the CPU and GPU are placed on the same die may be more elegant and natural, as both blocks benefit from characteristics offered by the most advanced CMOS nodes. That is, technology choice rather than access to DRAM, may be the better dominant constraint for partitioning decisions. A Split-Memory (sometimes called Hybrid Memory) solution may be an opportunity to find a global optimum. In this scheme, one memory—e.g. WideIO—would be dedicated to the GPU (cares about memory bandwidth), and a separate memory stack—e.g. LPDDR—would support the CPU (cares about latency), and the rest of the chip. Whereas this solution may seem attractive and may have good cost structure, it requires a new OS feature (Non-Uniform-Memory-Access

NUMA) and incremental features in Si (multiple controlers, concurrency logic…).

A different concept architecture where access to OS-controlled memory stack is the driving consideration for a partition may be an alternative option. With this concept the partition is driven by separating IP blocks into a group that requires access to OS managed memory, and a group that require memory, but not necessarily OS controlled. Then, all the blocks that require priority access to a memory that is managed by the OS would be placed on one die (e.g., CPU, GPU…), along with the controller, and all the blocks that require access to a (separate) memory that is managed by local software (not OS), such as Modem and Multimedia Frame Buffers, could be placed on the other die. The concept partition could then allocate 3 GB to Die 1 (GPU, CPU…) and 1 GB to Die2 (Modem, Frame Buffer…).

The pros and cons of this partition would then be as shown in Table 4.1.

With this architecture, there are then opportunities to reduce the overall SiP component cost by optimizing Si technology choices for each die. On the other hand there is some increase in Si area overhead, and unknown complexity in managing the coherency of the two separate memory stacks, etc. It is therefore a different tradeoff point.

ii. *Scalable architecture*: The SiP value propositions, such as Split Die concept, and various associated tradeoffs discussed (Chaps. 2 and 3) are focused on optimizing the manufacturing costs, rather than the of Non-Recurring Engineering (NRE) cost. An alternative concept architecture that focuses on the NRE benefits, rather than just the AUC benefits, may be attractive. Fundamentally, this concept is to break up a mono-die SoC such that the overall component performance (not functionality) is scaled with addition of incremental Si area. An example could be a partition where a single die meets the requirements of a mid-tier phone, and a two-die stack is competitive for

Table 4.1 Split system memory architecture—a summary of the key pro's and con'sfor the split memory architecture proposed in the text

Pro	Con
Avoid NUMA requirement	Extra Si area required for the second memory controller
Maintain modem performance with a hybrid private memory	Managing coherency across the two separate memories
Preserve the flexibility of partitioning along the technology affinity lines	Potentially increase external I/O pin count (unless use 3D stacking w WideIO for private memory)
Minimize the D2D bandwidth, and thereby reduce the interposer technology cost	Potentially increase cost of DRAM (unless use lower cost performance LPDDR stack)

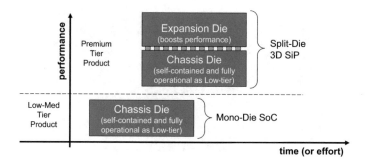

Fig. 4.1 Concept scalable architecture—an illustration of a concept architecture where stacking of additional tiers results in increased system performance (not functionality), showing an example case where one die ('chasses') targets a mid-range phone, and a 3D SiP stack (chasses + expansion die) targets high end phone

premium tier positioning. The concept is illustrated in the Fig. 4.1, showing a 'chasses' die targeting a given mid-tier level of performance, and an additional 'expansion die' to produce a higher tier level of performance.

This approach would incur area penalties versus two Mono-Die SoC, each optimized for a given target performance level. In addition, there would be area overhead associated with the interconnecting the die—i.e. the usual D2D overhead. But, on the other hand, a single design could be used to target multiple market tiers, thereby saving on the engineering costs.

iii. ***Distributed architecture***: It is obvious that the trends in the mobile market—including higher performance, more power and thinner and smaller phones—all exacerbate thermal management challenges. Another concept architecture, then could be to optimize–for–thermal, and to change the architecture from the current 'integrated architecture', characterized by a small number of large chips, to a 'distributed architecture', characterized by many small chips. This is illustrated in the sketch in Fig. 4.2. That is, with the integrated architecture, multiple processing units, performing much of the work during a given hi-power use case, are clustered together on a SoC and create a hot spot. With the distributed architecture, the display is segregated into a series of 'tiles' (not physically separate entities, but groups of addresses on the display). Then each tile is driven by a small chip containing a processor, driver, PMIC, etc., functionality, so that the workload—and hence heat generation—is distributed across multiple die resulting in less intense hot spots. Whereas this approach could clearly have significant power, power density and thermal management advantages, it may be prohibitively expensive.

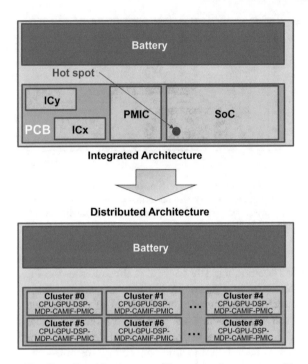

Fig. 4.2 Concept distributed architecture—an illustration of a concept architecture where the system is implemented using many small chips, as opposed to the current integrated architecture, where the system is implemented using a small number of large chips

4.2.2 Physical Partitioning Concepts

All the architecture concepts that have been discussed, e.g. the area-based Split Die or the Memory-on-Logic Stacked Die, are examples of a common partitioning 'philosophy'. That is, so far all the concepts used a "*Functional Partitioning*" philosophy, where the functionality of the lower level IP blocks is unchanged by the partitioning. Thus, for 2.5D Split Die partition and/or WideIO Memory-on-Logic stack, it is assumed that the basic functionality of the IP blocks is maintained through the partitioning, without exploring the possibilities of re-designing the IP blocks or changing the distribution of macros across a chipset. This 'functional partitioning' philosophy is the most obvious and easily intuitive approach.

Some different possible partitioning philosophies are briefly outlined below.

i. *'Functional Partition' Philosophy* assumes minimal disruption to existing
   architectures, and the current definition of IP blocks remains unchanged, as
   mentioned. The only thing that is changed is which IP resides on which die,
   and how these are interconnected. Hence this type of partition goes well with
   integration technologies that enable just the interconnect wires—such as for
   example the current 2.5D or 3D technology options. And, this partition goes
   well with leveraging the existing 2D design environment with minor
   enhancements to enable SiP design.

ii. *'Feature Partition' Philosophy* assumes that the existing functions may be
    reallocated across an existing chipset—for example analog blocks (DAC,
    ADC.) may be moved from the Application Processor die to a PMIC die.
    However, this approach may require that some macros be broken up—because
    some features may need to remain in close physical proximity to the
    Application Processor die. For example, R-C network for PDN management,
    thermometers for thermal management, etc. all require to be physically near
    the App Processor. This kind of a feature re-allocation might be forced by
    limitations of an advanced CMOS technology to support analog functions, and
    therefore requires 2.5D or 3D integration technology that includes more than
    just the interconnect wires. For example, this philosophy could be enabled by
    use of Si interposers that include passive elements (e.g. R-C-L for PDN
    management), or diodes (e.g., thermometer for thermal management), in
    addition to wires. And, this partition philosophy probably requires a polygon
    level custom design environment that is 2.5D and/or 3D aware.

iii. *'IP partition' Philosophy* assumes that some of the system functions and/or IP
     blocks may need to be entirely redefined—for example, because the advanced
     CMOS technology might not be able to support high voltage IOs. This par-
     tition would then require a very fine grain integration technology that
     approximates the mono-die, and would be enabled by, for example use of
     2.5D Si interposers that include active elements such as analog transistors
     (e.g., for I/Os and/or ESD protection), or even digital transistors (e.g., for chip
     utilities such as clocks or power gating circuits), or a heterogeneous 3D
     integration technology. And the design environment would need to be fully
     upgraded to support multi-die 3D design. Note that the PHY-Die Split,
     described in Chap. 2, is an example of this type of partition philosophy.

These partitioning Philosophies and the corresponding fundamental CMOS
technology constraints and Integration technology opportunities are summarized in
Table 4.2.

Table 4.2 Partitioning philosophies—a summary of various 'Philosophies' that can be used for selecting a suitable architectural partition for 2.5D or 3D SiP, and the corresponding fundamental technology constraints and Integration opportunities

Concept		Constraint	Vehicle
2D	Mono-die SoC	***Monodie Si* = Continue System integration through More Moore CMOS scaling** • Assume CMOS More-Moore continues… − no new disruptive constraints on Si devices − Cost-performance-power curve carries on	Monodie SoC Substrate Package
2.5D	2.5D interposer	***Functional partition* = Split Die with existing subsystem blocks assigned to a given die** • Assume CMOS has no new constraints • Must enable D2D interconnect for required BW • Minimize disruption to the architecture and design	Die 1 Die 2 Interposer with wires
3D	3D interposer	***Feature partition* = enable "near" subfunction with system blocks assigned to suitable chip-set die** • Assume CMOS cannot support (some) analog • Must enable physical proximity	Die 1 Interposer with wires + more ?
3D	3D LoL	***IP partition* = enable tight interface between system or SoC functions assigned to suitable die** • Assume CMOS has constraints (e.g. no HV I/Os) • Must enable interface between IP on different die • Disruption to digital design environment	Die 2 Die 3 3D Logic-on-Logic Stack

4.3   More-than-Moore Design for Multi-Physics Opportunities

2.5D, and especially 3D, technologies place multiple die in intimate contact with each other, thereby precipitating opportunities for interactions among the die, in all physical domains: electrical, thermal, and mechanical. This section is focused on the physics of the interactions and the associated modeling methodologies—especially in the thermal and mechanical stress domains. The insights derived from deploying an eco-system for characterizing and modeling these interactions are also summarized. Note that the next Sect. 4.4 focuses on definition of methodologies for importing these interactions onto a chip design flow, and the associated EDA tools.

4.3.1 Challenges

Dealing with multi-physics is a new challenge to IC design—for the incumbent 2D SoC products as much as for 2.5D/3D SiP designs. Traditionally, thermal and mechanical issues were handled by system designers addressing the overall box-level design, who would then normally precipitate requirements and/or constraints for package design. Rarely did these considerations trickle down to impact Si design. Some global guidelines dictated general wisdom—such as "low power is good", "certain packages have certain maximum die sizes", "corner pins should be used for P/G connections only", etc. The challenges of designing a modern mono-die SoC, just in electrical domain are daunting, so that multi die interactions, and especially thermal and mechanical considerations, are always abstracted out, and normally managed by package and/or system designers.

However, thermal issues are the fundamental limitation of CMOS technologies, and managing mechanical stress will be the next big barrier. The challenges can no longer be addressed purely by package design—especially so for the 2.5D/3D technologies, since in addition to the intrinsic trends associated with increased power density and decreased die thickness there is also the incremental multi-physics interactions that must be considered. That is, 2.5D and 3D SiP technologies exacerbate the intrinsic thermal and mechanical trends and are a harbinger of things to come in 2D SoCs. A paradigm shift is required—one which will ultimately pull thermal and mechanical considerations into Si design methodology, to enable co-optimization across multi-physical and multidisciplinary domains.

The underlying reason for the paradigm shift is pretty much inevitable consequence of CMOS scaling. As technologies scale, the number of devices per unit area, the performance per device, and the number of devices per chip all increase—all leading to higher power and power density. This is illustrated in Fig. 4.3,

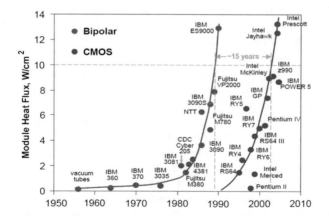

Fig. 4.3 Power density trends—a chart of heat density versus calendar time trends for high end CPU die in Bipolar ECL and CMOS technologies (Lucent), and highlighting that CMOS products are lagging ECL products in terms heat flux by about 15 years

showing that CMOS products are lagging ECL products in terms of heat density by a decade or two—but are following a similar exponential growth curve. BiPolar ECL technologies were eventually abandoned circa 1990, due to the prohibitive rise in the power density, and the associated cooling challenges. 20+ years later high end CMOS devices have reached comparable power densities. Higher power and power density precipitate thermal challenges, and at some point, no cooling solution can deal with the heat generated. Furthermore, heat sinks, heat pipes, blowers, water cooling, etc. are deployed in the servers and PCs, but cannot be readily adopted for mobile applications. Hence, there is a need to import thermal awareness into the design flow—so that judicious tradeoff between performance, area and temperature versus cooling solution cost, etc., can be enabled.

For mechanical stress, a similar trend is driven by the reduction in system form factor, driving commensurate reduction in package and die thicknesses. With die thickness at 100 u or below, ubiquitous for mobile applications, IC's are susceptible to local bending and warping, resulting in significant shifts in performance. Hence, whereas in the past the stress effects on device performance were small, and could be accounted for through use of excess margins, in the future, with reduced performance slacks, there will be an increasing need to account for stress effects explicitly in the design flow.

Clearly, the ultimate objective is therefore to develop a universal co-design methodology and associated infrastructure that would allow modeling and simulation of the all the phenomena—electrical, thermal, and mechanical—so that chip, package, and system design can all be suitably optimized for cost (yield), power and performance, within the thermal and stress envelope, with minimum excess margin. Lofty goal, since:

i. *Multi-physics*: the phenomena are in electrical, thermal and mechanical domains, and these clearly affect each other. Electrical—Thermal interactions can be modeled as Joule heating, but the transistor response to temperature is complex, dependent on transistor architecture and type, susceptible to thermal runaway, etc. Electrical—Mechanical interactions are modeled as Piezoelectric effects, but these are extremely complicated and dependent on materials, crystal and device orientations, bias conditions, circuit design, etc. Thermal–Mechanical interactions are modeled by various Stress–Strain models, with complex dependencies on basic material characteristics (CTE, Young's Modulus...), which, in some cases, also vary with temperature, etc. That is, modeling the interactions between any two domains is hard enough. Modeling dependence and interactions between all three physical domains is extremely hard. And modeling these interactions across all scales that matter to any one of the mechanisms is neigh well impossible. Hence for modeling purposes, the interactions must somehow be confined, the physics must be simplified, and the problem to be simulated must be suitably framed. Not easy!

ii. *Multi-scale*: multi-physics modeling is mostly done using some version of FEM (Finite Element Method) or CFD (Computational Fluid Dynamics) techniques. These techniques require some form of meshing—to break down the phenomena into elements small enough to be modeled. However, meshing techniques are challenged by different scales. Thus, chip interactions must be described at $\sim$nm level (to comprehend individual transistors), package interactions must be modeled at ~ 10 um to ~ 1 mm level (to account for features like uBumps and on up), and system phenomena must be described at ~ 1–100 mm scale (to account for system features like heat spreaders, frames, etc.). Therefore, a single mesh model would have to span many orders of magnitude. Similarly, with modeling in time domain. Electrical phenomena occur in $\sim$ns, but some of the thermal, and hence mechanical, responses take sec's to min's, or, even years if reliability effects (stress- and electromigration…) are modeled as well. Simulating interactions across these types of range of scale is not practical with mesh based solutions, so that some kind of sub-modeling, or compact behavioral modeling, or response surface modeling, etc., is required. Not easy!

iii. ***Multi-schedule***: the phenomena have chip, package, and system dependencies, so that good modeling should lead to global optimization across all three of these 'disciplines'. However, normally, the designs are done sequentially and hierarchically, i.e., chip design is done first (takes ~ 1–2 years), then package design (takes ~ 3–6 months), and then system design (another ~ 6–12 months), with each phase dictating the constraints for the next one. This sequential design flow makes global optimization impossible, and either some kind of default format for bidirectional exchange of constraint and performance tradeoffs, or synchronous co-design of all elements, or iterative methodology, or…, is required. Not easy!

Thus, modeling 2.5D and 3D SiP designs—involving multiple die and/or thin double-sided die and/or interposers and/or new features like TSV and uBump, at chip, package and system level in electrical, thermal and/or mechanical domains—cannot be done explicitly. Shortcuts and tricks must be used.

4.3.2   Infrastructure Requirements

Thermal and mechanical modeling requires information about material properties that are new to the Si design eco-system. Electrical properties required for Si design are routinely characterized by the supply chain partners, leveraging pretty much standard characterization, model extraction and QA practices, and are described through elegant compact models like SPICE and standardized PDKs. There is nothing comparable for thermal and mechanical modeling anywhere. The OSATS do provide—actually mostly pass through—the packaging material characteristics supplied by the raw materials vendors. However, these are derived on bulk material samples, and as such may not be representative of the characteristics of the thin film and/or microfeature implementation. In addition, the characterization techniques and test points are not standardized, and tend to vary from supplier to supplier, making comparisons and/or extrapolations to desired conditions, hard to do. Furthermore, material properties of the input chemicals do not necessarily represent the characteristics of films at the end of a manufacturing line—due to the restructuring that may occur through the manufacturing process. Hence, models based on the bulk material properties are suspect and need to be validated for every implementation—thereby limiting the usefulness of the models. In addition—especially in case of mechanical stress—there are residual properties that are specific to a process technology, i.e., residual stress in the film stack is a function of process thermal history, stacking flow, etc.

Hence, in addition to developing a modeling and simulation methodology, there is a need to develop standard material characterization practices and standard formats across the industry—for parameters like thermal conductivity, CTE, Tg, Young's Modulus and Poisson ratio, moisture absorption, residual stress, etc. These parameters—being process specific—should be owned by the manufacturing entity (OSATs, foundries...) and supplied to the user community—much as is the case with foundry-supplied SPICE models and PDKs describing Si device electrical characteristics for a given technology. That is, ideally, SPICE-like compact, behavioral models, provided and calibrated by the supply chain partners, should be enabled to optimize SiP products. It took the industry several decades to deploy SPICE—with all the accompanying infrastructure and standards—that make it now such a powerful tool. It will likely also take years to bring up comparable infrastructure and capability for thermal and mechanical simulations. However, for disruptive MtM class of technology solutions—such as the 2.5D and 3D integration technologies—this capability is a critical enabling requirement.

Hence, an industrywide effort to develop a methodology, and accompanying infrastructure, for thermal and mechanical modeling, that is suitable for a fabless user, and that would interface in real time with product design is required. The effort should be focused on obtaining the necessary material and residual stress data, and on establishing industry practices to provide this on routine bases. This effort involves collaboration between foundries, OSATS, universities, Research Institutes and Standards bodies to evolve a complete practice for supporting and supplying this type of data. Whereas it is always possible to do almost anything manually and on one-off custom bases, in order for high volume commercial products to routinely leverage 2.5D and 3D technologies, this infrastructure must be there.

4.3.3   Thermal Management

The cornerstone of thermal mitigation is thermal modeling—after all, phenomena that are not quantified cannot be fixed and/or optimized. The reality of thermal modeling is that it is far more complex than the traditional oversimplified relationship describing junction temperature as:

$$T_j = P * \theta_{j-a} + T_a$$

where

T_j = Junction Temperature
P = Power Dissipation
Q_{j-a} = Thermal Resistance of a package
T_a = Ambient Temperature

In fact, thermal modeling is based on the physics of heat flow, and as such it is a multidimensional challenge where local behavior cannot be modeled in the absence of various global considerations. Heat must flow down a temperature gradient to somewhere, and eventually be absorbed by the environment. Hence, thermal modeling must encompass the entire heat path, from the hot spot where it is generated—such as a region on a chip—to the bulk of Si, to package, to PCB and eventually to the entire system and the environment. Heat flow, and the temperature distribution along its path, can therefore be dominated by any feature in the path. Thus, temperature of a hot spot on the chip can be determined by the thermal resistance of the package or the board or even the ability of the system to dissipate heat to the environment. For example, with relatively small devices, such as phones, steady state power is in fact ultimately limited to ~ 4–5 W by the system's ability to dissipate heat to the environment. However, since the heat wave moves through the system relatively slowly, it takes several minutes to reach steady state condition in a mobile device. Therefore, short bursts of power dissipation—of the order of seconds—can result in thermal transients where the local peak temperature is dominated by the local characteristics, such as, for example chip geometry (Si die acts as a heat spreader so that big, fat die is better than small, skinny one). Thus, in mobile devices, power bursts can produce local chip hot spots that can exceed maximum rated Si junction temperature, and/or steady state power dissipation can exceed the maximum rated skin temperature. Therefore, all time–domain combinations, that encompass all possible distribution of power bursts, should be modeled. Finally, thermal modeling is only as good as the input parameters: principally power and material properties. Power and power density is use-case dependent, and needs to be modeled at granularity of, at least, IP level. Furthermore, power is in fact a distribution that varies with process device parameters (fast vs slow process corners), so that it too should be represented in form of corner values. Material characteristics are also a distribution that encompass process and material variability, and should also have corner model values. Therefore, physically correct thermal modeling that includes all these considerations and variables, is complicated—beyond the generic multiscale challenges. In fact, in practice it is too complex, so that thermal modeling is typically based on 'nominal' Power and Material properties, and includes steady state and a few use case specific transient simulations.

With modern high end SoC's thermal limits are readily exceeded, impacting performance, reliability, and ultimately functionality, of a device, and in some extreme cases resulting in material destruction or even outright fires. Consequently, the current standard practice, is to include thermal sensors in the chip, and/or system, to monitor the various local temperatures, and to 'throttle down' chip performance—typically by turning down the clock frequencies and/or rail voltages—when critical temperatures are reached. For phones the critical temperatures are typically ~ 100 °C for Si junctions and ~ 45 °C for phone skin temperature. In this context, design decisions include the placement of the temperatures sensors on the chip, and the management of the various necessary guard bands. Thermal modeling is still required to guide these decisions.

In addition, internal thermal interactions—and specifically managing the effect of hot spots on one die on peak temperature of another die—are often perceived to be one of the fundamental limitations of 2.5D and especially 3D SiPs. Interestingly, as described in Chap. 3, a system design option may dominate the mitigation of internal thermal interactions.

All the above considerations are equally applicable for 2D SoC's as for 2.5D/3D SiP's—except that the construction of the SiP components is typically more complex, requiring modeling of more layers and more heat sources, thereby making thermal simulations that much more challenging. The intent here is to outline the options and opportunities for generic Design-for-Thermal practices for 2.5D/3D SiP design. Note that the 3 phases of design—PathFinding, TechTuning and Design Authoring—are more fully defined in Sect. 4.4.

i. *Design Authoring for Thermal*: Several different thermal simulation methodologies and tools have been developed to address the thermal concerns specific to Si chip design, ranging from the general purpose thermal simulators (e.g. Mentor FloTherm, Ansys IcePack) to tools specialized for use within Si design flow (e.g. Cadence QRC, Synopsys (Gradient), Silvaco (Invarian)…).

In general, if specialized tools are used within an IC design flow, their accuracy is compromised by the inaccuracy of the input constraints—such as how the package and the rest of the system is modeled. That is, typically, the abbreviations used to model the package/PCB/System behavior dominate the accuracy of the thermal predictions at Si chip level, thereby undermining their credibility and usefulness to directly affect chip design decisions.

On the other hand, full CFD/FEM tools can be used to describe accurately the package and system behavior, but are typically too slow to be useful as an interactive input within actual Si design flow. Furthermore, the details required for input to these tools (specific device construction, specific use case, power and power distribution, environment, interface to environment, etc.) typically do not exist while Si design is done—forcing use of some kind of a default reference system. But, when a given reference system is used as a bases for chip design, then the chip is signed off for a single specific package and/or system. Thus, with either approach—general purpose or specialized thermal tools—

thermal simulations typically end up producing a relative Figure of Merit metric, rather than a signoff constraint that affects actual logic design or chip floorplan. The real issue, even if the infrastructure was implemented and the effects of the package and system accurately portrayed, is that the tradeoff between thermal performance and other constraints—such as electrical performance or cost—is typically not well defined. That is, how many $$ AUC or MHz clock frequency is worth how many °C temperature? This is very difficult to specify explicitly. Typically, the best and most practical way of addressing this conundrum is to use fixed benchmarking use-case scores as a design objective (e.g. AnTuTu score in mobile space). Note however that this score is as much, if not more, dependent on system design parameters than on chip design specifics.

Thus, in practice, design-for-thermal methodology at Si chip Design Authoring phase is difficult to implement. Furthermore, the thermal mitigation practices—such as thermally driven performance throttling currently ubiquitously used in SoC design—somewhat obviate the need for detailed and interactive thermal simulations within a chip design flow.

ii. *TechTuning for Thermal*: Given the observations outlined above, the objective of TechTuning for Thermal is less about creating layout and floorplanning rules for design authoring, and more about generating specifications for the target process technology. These can be used to optimize the selection of materials and packages construction. For example, the architecture if the package for WideIO M-o-L can be driven by thermal considerations to include a requirement for zero overmold thickness in order to couple the die stack to a heat spreader.

Nevertheless, there are of course opportunities to optimize package and die design for thermal considerations, and to leverage thermal design rules and guidelines. Most of these opportunities can be encompassed within a set of generic guidelines for SiP design—more as a general quality-of-design practice, rather than a specific design rule. Some examples of these include:

- Don't place high power elements in die corners—prefer die center
- Don't place high power elements close together—spread them out if possible
- Don't place thermally sensitive circuitry next to a hot spot—offset as much as possible
- Constrain max power density for a given package class—or use a thermally better package
- Thermal TSV vias have minimal effect on thermal performance at reasonable power densities
- Thermal vias in package substrate and/or thermal bumps for chip-package attach do haves positive effect on thermal performance, but the impact saturates with via count.

iii. ***PathFinding for Thermal***—The biggest 'knob' for thermal mitigation is at the architecture level—and hence should be considered as a part of the tradeoffs analyzed in the PathFinding flow. That is, whereas Design-for-Thermal at chip level is a Figure of Merit metric, at Architecture level it is a real design knob that can drive real architectural decsions. However, as outlined above, the usual thermal modeling methodologies typically require a lot of detailed inputs (not available during pathfinding phase), are optimized for accuracy, and are too cumbersome to simulate all the candidate variants that are considered during this phase of design. Consequently, a PathFinding for Thermal methodology is required, and has been proposed (Radojcic 2014). The basic elements of the proposed methodology are outlined below:

- A set of simplified 'concept test vehicles' are used for thermal simulations—rather than a detailed reference system. Different test vehicles are used to reflect different thermal constraints. Thus, the test vehicle for peak junction temperature is defined to represent chip and package features (die and package geometries, hot spot power density and location, etc.), while a separate test vehicle, representing PCB and system features (PCB and handset geometries, display and skin thickness, etc.), is used to address skin temperature constraints. These vehicles represent constraints which are separated in physical domain (um-mm scale vs mm-cm scale, respectively) and in the time-domain (seconds for junction temperature, minutes for skin temperature). The use of separate vehicles addresses the multi-scale modeling challenges.

- The test vehicles are defined to represent a given architectural feature and its sensitivity to deviation from ideal. For example, 'ideal' system vehicle assumes that the PCB is isothermal and that the heat spreading across it is perfect, and a 'deviation' case assumes a hot spot on the PCB of certain size and density. Similarly, 'ideal' component level vehicle assumes that the power is perfectly uniformly spread across the die, and a 'deviation' case introduces a die level hot spot of certain density and size. The thermal simulations on these vehicles under these assumptions then represent sensitivity of a certain architecture to deviations from idealized design—and as such is an approximate measure of thermal robustness.

- In order to manage the tradeoff between thermal and other design objectives, thermal simulations are inverted and temperature is treated as an Input Constraint rather than an Output Variable. That is, since power is the common denominator for performance and temperature, thermal simulations are performed to determine the max power for a given target peak temperature. For a phone, or a tablet, the typical thermal constraints are max skin temperature (40–45 °C) and max IC junction temperature (80–125 °C). Thermal simulations are done to define the power envelope required to stay within these thermal constraints.

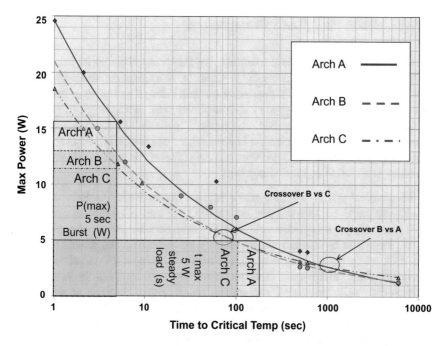

Fig. 4.4 PathFinding for architecture for thermal—a plot of max power versus time-to-critical temperature, for concept test vehicles representing key attributes of three chip-set architectures (3D M-o-L SiP, PoP, 2 separate packages), highlighting the relative thermal merits at arbitrary comparative points (5 s burst of power, and constant 5 W power load)

The output from the proposed PathFinding for Thermal methodology is illustrated in Fig. 4.4, showing a plot of max power versus time-to throttle (i.e. time to reach specified Max temperature), for 3 concept architectures (e.g. 3D SiP, PoP, multiple 2D components...). Also highlighted are the relative thermal merits of the 3 architectures at two arbitrary comparative points; corresponding to a 5 sec burst of power, and to a constant 5 W power load. In the example case illustrated, 'Architecture A' is superior for transient power bursts, but Architecture B is better under steady state conditions.

Thus, with the use of these simplified constructs, and the proposed methodology, stimulations are sufficiently simplified to address multiple design cases, and the results provide insights into the *relative* strengths and weaknesses of each architecture. That is, as in the example case illustrated, analyses is used to measure the relative thermal margin of an architecture, rather than to estimate an actual absolute max power limit. Note that the Thermal methodology described here is also a demonstration of a generic PathFinding methodology per-se, i.e. an approach for simplifying a problem to the point where modeling and simulations can be performed with reasonable ease and speed, by compromising accuracy, and extracting insights with fidelity for architecture level design.

4.3.4 Mechanical Stress Management

Mechanical Stress interactions are dominated by the local, rather than global factors—unlike thermal. As discussed briefly in Chap. 3, managing the stress–strain phenomena is a complex tradeoff, dependent on many design and process parameters. Hence a practical methodology to model the phenomena, so that any sensitivities can be designed out, is also required. A proposed methodology has been described elsewhere (Tengfei et al. 2015; Xu and Karmarkar 2011 and Sukharev et al. 2010 in Chap. 3; Papanikolaou et al. 2011; Wu et al. 2011; Zschech et al. 2011; Ho et al. 2014; Sukharev et al. 2016) and is outlined here.

i. *Mechanical Stress Driven Failure Modes*: Stress–Strain phenomena are not exactly new, or reserved for advanced 2.5D and 3D technology options. Various Chip-Package-Interactions (CPI) have been managed for decades, and are known to be exacerbated by industry adoption of softer dielectrics in the BEOL stack (lower k dielectrics), use of harder chip attach materials (Pb free solder, Cu…) and larger/thinner die. Similarly, managing on-die stress issues, including stress migration and various stress driven dielectric cracking phenomena, has been a challenge for a while—exacerbated by adoption of aggressive strain engineering tricks in transistor architecture, and use of harder conductors and softer dielectrics within BEOL stack.

   However, some of the *symptoms* of stress are novel. Traditional concern was with cracks, die chips, delamination, creep, filaments, deformation, warpage, etc.—i.e. some form of physical distortion that various stress driven phenomena precipitated. And whereas these are obviously serious issues that must be managed, the focus here is on the stress driven phenomena that cause shifts in the *electrical performance*. The only practical approach for addressing these is to proactively design them out—hence the emphasis here on eCPI - as mentioned in Chap. 3. Furthermore, more complex 2.5D/3D SiP packages can precipitate multiple stress interactions—for example different die interacting with each other, with different stacking overlaps and underlaps, or different alignment between topside uBumps versus bottom side Cu pillars, and with very thin FanOut kind of packages versus BGA balls, etc. Hence any 2.5D/3D SiP design methodology must especially focus on designing stress effects out. Some general definitions:

   - *mCPI* (**m**echanical Chip-Package-Interaction): as defined in Chap. 3, result in cracks, delamination, fatigue, creep, etc., i.e. defects that can be seen and physically detected. Typically, mCPI results in open/short type of fails. The driving mechanism is CTE mismatch between the package and the Si die. An example of mitigation practices is the constraints on Distance to Neutral Point (DNP) that are sometimes applied as design rules for managing placement of signal pins on a large die and/or package.
   - *eCPI* (**e**lectrical Chip-Package-Interactions): results in shifts in carrier mobility, and hence in device performance variability. Mobility shifts of the

order of $\sim 10\%$ can be precipitated at stress levels of the order of high 10s MPa or low 100s MPa's, resulting in transistor Idd shifts of the order of 5–10%. This is a big deal for modern CMOS devices. The interaction does not cause physical deformation that can be seen—no "smoking hole"—and is hence very difficult to fault isolate. eCPI is driven by CTE mismatch between the chip and the organic mold, underfil and/or substrate materials resulting in die edge/corners being bent, and/or volumetric shrinkage of the underfill materials, resulting in the chip being pulled towards the package, and the ball pushing on Si. An example is parametric yield loss sometimes observed with high performance DAC circuits when chip attach balls are placed on top of the DACs. Note that the principal source of stress is a time-zero problem (at the end-of-line caused by cooling from processing temperature to room temperature)—same or similar as for mCPI, but that the sensitivity to stress may be a time-dependent reliability issue, as transistors age (due to unrelated phenomena).

- **eCBI** (electrical Chip-Board-Interactions): results in shift in carrier mobility and device performance, identical as with eCPI, but due to chip interacting with the PCB (in addition to the package). eCBI is especially a concern with the FO class of packages. With this technology, the package is very flimsy, and mechanical interactions, driven by CTE mismatch and/or underfill shrinkage, are between the Si die and the PCB. Note that whereas the mechanism is identical to eCPI, the symptom has an incremental appearance of a reliability issue, since failures are observed only after attach of a component onto the client board, rather than during manufacturer's testing. That is, eCPI and eCBI are non-catastrophic stress driven phenomena that can result in increased and/or inconsistent yield loss, and/or field returns—very hard to fault isolate, identify, and therefore fix. Note that due to the 'soft' nature of the eCPI/eCBI fails, a comprehensive electrical test is required for verification and characterization of the fails. Furthermore, the test may have reproducibility issues that are affected by test set up, temperature, test jig, etc. All this makes identification and verification of the fails difficult, and the definition of corrective action, especially with the distributed supply chain, nearly impossible. Hence eCPI/eCBI has to be addressed proactively through modeling and design; else the reactive corrective action cycle is long, costly and unpredictable. Note that the mCPI type of physical effects are typically triggered at stress levels of the order of few 100s of MPa, but that for thin die ($\sim >100$ um) in current CMOS technologies, the eCPI effects (appreciable carrier mobility shift), can take place at high 10s of MPa, i.e. the eCPI effects can be more sensitive to stress then the traditional mCPI effects.

Since eCPI/eCBI phenomena are exacerbated by the wafer thinness, they are of especially a concern for 2.5D and 3D SiP's, i.e. thin mono-die 2D SoC's are equally susceptible to eCPI phenomena if the die are thin. Figure 4.5 highlights the current 2D SoC thickness range for laptop/tablet applications, mobile devices and target 3D SiP implementations.

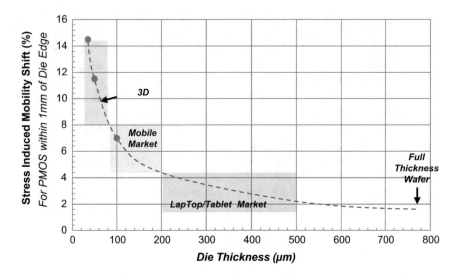

Fig. 4.5 Trend in die thickness—a plot of strain induced mobility shift versus die thickness, for a given concept implementation (constant die size, package, material set, etc.), showing a strong dependence on die thinness, and highlighting the typical current range for various application markets

ii. *Failure Mechanism and Dependencies*: The fundamental piezoelectric mechanism in Si has been known for decades. Deformation of the Si crystal lattice—caused by any form of mechanical stress—alters the local distances between the atoms in the lattice, and results in change in the electronic band structure leading to a shift in carrier mobility, due to the change of the slopes of the E-k curves, which lead to change in the carrier effective mass. As such, the change is a function of the direction of the stress vector, relative to the crystal lattice. Hence sensitivity of carrier mobility to stress varies with stress direction and crystal orientation (different minima in the E-k curve respond differently to stress). However, managing the effect in practice is far more complicated with many multi-domain dependencies. Examples of the key variables that directly affect the eCPI phenomena:

- Variables that affect the CPI source of Stress:

  - Die Geometry
  - BEOL stack up
  - Si Stress Free Temperature
  - On Die Layout features—e.g. wide metal lines
  - Package Construction and Geometries
  - Material properties (Young's Modulus, Poisson Ratio, Tg…)
  - Assembly Process Temperatures
  - Placement of Die in the Package
  - Placement of Overlap Edges (for 3D stacked die)

  – Placement of Die Bumps (and Backside uBumps for 3D)
  – Board Level Underfill Characteristics

• Variables that affect the sensitivity to strain:

  – Crystal Orientation, i.e. <111> or <100> or...
  – Device Orientation versus Wafer Flat
  – Device Architecture (Planar vs FinFET)
  – Device Type (PMOS vs. NMOS)
  – Device Size (L and W)
  – Nature of IP design (e.g. DAC, memory array...)
  – Chip Floorplan, i.e. the placement of a given IP
  – Operating Conditions, i.e. lowest ambient Temp...

Clearly a very complex phenomenon with many dependencies and very difficult to model explicitly, as illustrated in Fig. 4.6. Of course, given material properties and thermal history, stress can be described well with standard FEM type of modeling using standard FEM tools (Ansys, Abaqus...). However, these

Fig. 4.6 Effect of strain on carrier mobility—a cartoon illustration of stress vectors with molded Si die (**a**) and underfilled chip attach bump (**b**), highlighting the CTE mismatch driving the stress vectors, and corresponding map of electron and hole mobility shift

Cause: Mold⇔Si CTE Mismatch
Effect : Δu at die edges/corners

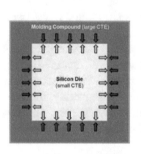

Cause: Underfil⇔Bump CTE
Effect : Δu around bumps

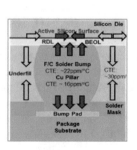

x: -200µm

tools typically do not include piezoelectric models required to describe the electrical response to strain, and specialized tools, along with technology specific piezoelectric coefficients, are required (e.g. Synopsys Sentaurus). Furthermore, the range of scales involved (from mm's for package features that source the stress down to nm's for transistor features that sense the stress) makes use of standard FEM methodology very challenging. Therefore, modeling the eCPI phenomena requires a new methodology, rather than just a right tool.

iii. ***Characterization and Modeling Methodology***: The general modeling and simulation methodology for addressing eCPI phenomena has been proposed and described elsewhere (Papanikolaou et al. 2011; Wu et al. 2011; Zschech et al. 2011; Ho et al. 2014; Sukharev et al. 2016), and since it is believed that managing the stress effects is of critical importance for commercial success of potential 2.5D/3D SiP products, is outlined here. The established current state of the art typically models 'in line' mechanical stress effects—such as warpage —and requires detailed process flow information. This is a TCAD class of capability supported by existing tools and methods. The proposed methodology focuses on defining an eco-system required to deploy a "DfM" class of capability that would model eCPI at the end of manufacturing line, at time-zero of operating life. The effect of time through field use, required to model the reliability effects ("DfR") as well as modeling all the interactions ("Multi-Physics"), are not addressed (yet?).

Stress/Strain effects—unlike thermal flows—are local, and dependent on the detailed polygon-level layout. Hence these phenomena are not suitable for addressing through a PathFinding flow—such as the one described above for thermal effects. Consequently, the focus is to develop a TechTuning class of a flow to generate layout rules and models that would then constrain Design Authoring. The fundamental elements of the methodology include:

- A simulation flow based on conventional FEM methodologies and tools (ag Ansys) to describe the stresses generated by the package. This flow uses meshing compatible with package dimensions (mm's) with sub-models compatible with BGA and C4 dimensions (um's). The output of this simulation is basically a contour map of displacement caused by CTE mismatch between the package layers and Si die—with global resolution to model displacement at the die level, and local resolution to model displacement around bumps.
- A simulation flow based on specialized methodology that comprehend the piezoelectric effect. This can be either a calculator that includes piezocoefficients (e.g. Matlab) to model global effects, or a mesh based tool with dimensions compatible with transistor dimensions (e.g. Synopsys Sentaurus) to model local effects. The output of this simulation is a contour map of mobility shifts overlaid over the die map.

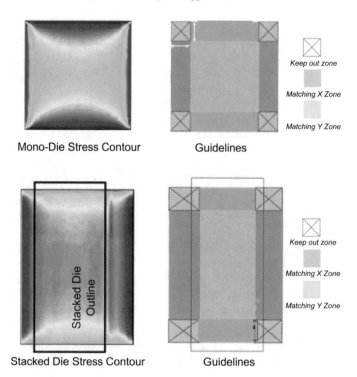

Mono-Die Stress Contour   Guidelines

Keep out zone

Matching X Zone

Matching Y Zone

Stacked Die Stress Contour   Guidelines

Keep out zone

Matching X Zone

Matching Y Zone

Fig. 4.7 eCPI floorplanning placement guidelines—a map of stress induced mobility shifts for a concept packaged mono-die SoC and stacked die SiP and an illustration of keep out zone (KoZ) guidelines for a given specific IP placement

- A set of Figure-of-Merit (FoM) circuits whose performance is characterized in SPICE as a function of mobility shifts. The FoM Circuits are selected for specific type of sensitivity to stress (e.g. DACs may be sensitive to stress gradients, latches may be sensitive to absolute stress, etc.). The output at this stage is a set of placement rules and guidelines for location of IP on a given die floorplan for a given target package type, as illustrated in Fig. 4.7. These general guidelines are then used in Design Authoring flow, as a good design practice. Note that the guidelines are specific for a given Si technology (driving the piezo coefficients), IP (driving the sensitivity to mobility shifts), and package (driving the magnitude of stress).

- A product signoff procedure to validate a specific design for sensitivity to eCPI effects. This is required because the eCPI effect is too complex to be generalized, so that the placement rules and guidelines are good proactive practices, but not a guarantee for any specific implementation. A product design can then be signed off by taking a specific package/die floorplan/IP combination through the steps outlined above, or by deploying a specialized checker (Mentor Caliber SSI) that includes a compact model for describing the stress sensitivities (Sukharev et al. 2016) (see Sect. 4.4 for further detail).

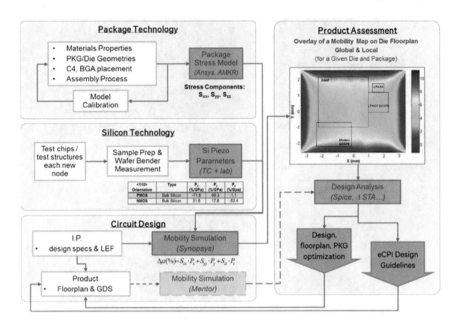

Fig. 4.8 eCPI management flow—a concept flow diagram illustrating a (possible) holistic implementation flow for managing the eCPI phenomena highlighting the roles of engineering disciplines found in a typical semiconductor company organization and EDA vendors with proven suitable tools

Note that in order to produce credible predictions, all the models must be calibrated, i.e. displacement models are dependent on material characteristics and thermal history, piezoelectric models are dependent on specific Si technology characteristics, IP models are dependent on calibrated SPICE, etc. The infrastructure identified in Sect. 4.3.2 is required. Note that a considerable amount of experience and data has been accumulated to date, so that the level of confidence in the methodology is actually quite high.

This is a very complex flow that is difficult to deploy, partially because it involves collaboration of different teams in a typical organization, working to different time scales, and bringing different skill mixes. The total implementation flow, and the corresponding typical organizational functions are illustrated in Fig. 4.8.

As illustrated, Package Technology team should obtain the material properties, calibrate the stress models, and simulate a target package. Si Technology team should obtain the piezocoefficients for the target CMOS technology (either by obtaining these from a foundry, or by implementing 'wafer bend' tests in lab to directly measure the coefficients). Circuit Design team should identify FoM IP and characterize it for stress sensitivity. And product design team should perform the product-specific assessment. Many moving parts and very complex—but, it is believed-the only way to ensure avoidance of eCPI fails. Note that the compact model based solution, identified in the figure as the "Mobility Simulation" box and further described in Sect. 4.4, in fact greatly simplifies the implementation flow (albeit the calibration and set up is still quite involved), and results in a 'normal' verification check in Design Authoring flow. Since eCPI is believed to be potentially an existential threat for 2.5D/3D SiPs, deploying some version of verification methodology for addressing the problem is important.

4.3.5   Electrical Interaction Management

Managing electrical interactions is clearly the most fundamental challenge for making successful 2.5D and/or 3D SiPs. The specific kinds of tradeoffs that 3D technologies precipitate have been described in Sect. 3.4.2. The overall design methodology—and the EDA tool strategy is described in Sect. 4.4. The focus in this section is on methodology for modeling and mitigating the interactions.

In general there is a choice of a strategy for dealing with the various new electrical interactions precipitated by the 2.5D/3D integration technologies:

i. **Restricted Layout + Lumped Models**: restrict placement and layout style, thereby greatly simplifying the modeling challenges. Thus, typically the TSV's are laid out in arrays or groups of arrays, TSV's are aligned directly with uBumps, BRDL pads and uBumps are a regular array, etc. In addition, a set of incremental placement restrictions and KoZ rules are used to manage potential interactions, e.g. an exclusion zone is defined around a TSV array to prohibit placement of sensitive analog circuitry close to the noise generating TSV array, or grounded guard rings are designed around the TSV array, or both. Use of these extreme layout restrictions allow use of lumped circuit models and SPICE simulations, as described in Chap. 3. That is, the unacceptable complexity of mesh based field solvers that would be required to model non-constrained features mandates limitations in the design flow. The 'restricted layout + lumped models' methodology is obviously easily integrated into standard design flows and practices, and no incremental EDA tools are needed.

   However, some cautionary comments:

   • Use of guard rings around sensitive circuitry may be of limited effectiveness when dealing TSV interactions. Guard rings are effective with planar structures where all the activity takes place in an x-y plane very close to the Si surface. With TSV's the noise is injected from the underside of the victim device, through the Si substrate, so that guard rings may not be effective. Use of grounded diffusion wells is much better—but comes at the cost of incremental mask layers.

- Use of lumped circuit models for SPICE simulation clearly requires that the models be calibrated for the target technology. Such calibration is typically only as good as the test chip used for it—i.e. the test chips must encompass all the layout restrictions and associated interactions contemplated for use in actual SiP product. In addition, the characterization and modeling methodology should be extended to cover the frequency domain of interest, using suitable S-Parameter type of test structures. Finally, Corner Models for addressing process, voltage and temperature variability in 2.5D/3D technologies are also required.
- Use of restricted layout flexibility, by definition, may erode the potential value proposition of 2.5D/3D design, since it may precipitate use of wire lengths that are longer than may be optimal. It is believed that it is quite adequate for some applications, such as M-o-L stacking; but it may be too restrictive for Logic-on-Logic stacking.

ii. *Flexible Design + Field Solver*. For some types of 2.5D/3D SiP, the layout restrictions that enable use of lumped models may be too restrictive. In those cases, a mesh based solver is required to model all effects and interactions, especially when the TSV is relatively large—such as the current Via-middle technologies. Note that, even if the stacking technology uses tiny TSV's (comparable in size to the normal BEOL vias), the parasitic effect is minimized, but not eliminated, because TSV's are embedded in semiconductive Si versus standard BEOL vias which are embedded in dielectric material. Mesh based field solvers are cumbersome and slow and typically not easily integrated in a standard design flow.

On the other hand, the advantages of the freedom to place TSV and/or uBump features anywhere in the floorplan can be compelling, resulting in shortest interconnect and most compact designs. A full 3D router would be required to manage this degree of freedom, i.e. manually placing TSV's and uBumps anywhere in the floorplan is not a practical solution when T2T interconnect involves several 1000's of wires. That is, this 'flexible design + field solver' approach requires a full 3D design environment, including a suitable mesh based 3D field solver.

It is possible that some mix of these two approaches could evolve, i.e. layout that goes beyond just the structured TSV arrays, but allows a number of different, pre-characterized, TSV constructs, is possible and may be an attractive tradeoff between requiring a full 3D design environment and limiting the layout flexibility.

4.4   More-than-Moore Design Methodology Opportunities

In most cases, disruptive technologies can be used in a product only if a product sourcing eco-system is available. A big part of that eco-system is design methodologies and EDA tools, since, ultimately, product designers experience process technologies only through the EDA tools and associated PDKs. Consequently, development of a suitable design environment is an essential enabler—and differentiator—for a disruptive More-than-Moore technology. The design environment requirements and the necessary incremental engineering needed to enable More-than-Moore technology options to intersect a real, commercial 2.5D or 3D SiP product are described in this section. The emphases here is on the EDA tools. Section 4.3 addressed Design-for-MultiPhysics.

4.4.1   Design Methodology Requirements—Ideal

2.5D and 3D integration technologies, by definition, bring (a) new degrees of freedom, and (b) new interactions. The new degrees of freedom must be leveraged to realize the value proposition—e.g. shorter wires leading to better power/performance... The new interactions must be modeled to design around them—e.g. TSV induced substrate noise. Thus, to design with the new technologies, EDA tools need to be upgraded to enable new degrees of freedom, and PDKs need to be expanded to model given 2.5D or 3D technology features. Furthermore, if a given MtM SiP product were to be designed with equal level of automation and productivity that is currently available for the design of a mono-die 2D SoC, then the entire design flow needs to be upgraded to manage the complexity without excessive manual interference.

In general, design of 2.5D and/or 3D SiPs requires EDA tools and design flows to support New Features and Multi-Die and Package Co-Design. Specifically, the tools need to understand the following:

- multiple technology files (PDKs) for multiple die + interposer + packaging technologies,
- new features like TSV, uBumps and BRDL,
- multiple die floorplans and package placements,
- multiple placement and routing styles,
- understand die that have two sides, and
- co-design Si, interposer and package.

And do all this simultaneously, i.e., not sequentially by leveraging 'one-die-at-a-time' approach.

In addition, with most MtM technologies and SiP products, multiple die are placed in intimate proximity and interact with each other in thermal, mechanical and/or electrical domains. These interactions must be imported into the design environment, ideally through use of in situ Multi-Physics models and simulators necessary to enable interactive optimization, or, practically, through use of layout restrictions and KoZ rules necessary to enable signoff. This was discussed in Sect. 4.3.

Much has been written about the specific requirements for 3D design environment (e.g., Xie et al. 2011; Papanikolaou et al. 2011; Wu et al. 2011; Radojcic

Table 4.3 Requirements for 2.5D and 3D SiP design—a table summary of the specific high level capabilities required for design of various target 2.5D and 3D integration schemes

New technology attributers/interactions versus mono-die SoC	2.0D	Hybrid technology	2.5D technology	3D technology	
	1 SoC-on-substrate	1 SoC-on-interposer-on-FO	2 split die-on interposer	M-o-L-on-substrate	L-o-L-on-substrate
New PDK features e.g., bumps, TSV's, uBumps	No	Yes	Yes	No	Yes
Multiple floorplans e.g., D1, D2, Interposer	No	No	Yes	No	Yes
Multiple P&R algorithms e.g., Manhattan versus anyway	No	Yes	Yes	No	No
Multiple technologies e.g., 10FF and 20SoC	No	No	Yes	No	Yes
Multiple sides per die e.g., backside features	No	No	No	No	Yes
Electrical interations e.g., D1 versus D2 performance	No	No	No	Yes	Yes
Thermal interactions e.g., D1 versus D2 hot spot	No	No	Yes	Yes	Yes
Mechanical interactions e.g., D1 versus D2 or FO	No	Yes	No	Yes	Yes

2008) and will not be discussed further here. However, the details vary with target technology as well as target application, and Table 4.3 illustrates the type of general design flow features that may be required to support design of a given 2.5D/3D SiP product type.

That is, developing a design environment for 2.5D/3D technologies involves a lot more than allowing a few extra layers in the GDS2 database, and representing new features in the physical layout domain. An approach for suitably representing each feature offered by the new technology paradigms at higher levels in the flow, including the methodologies for abstracting and margining the physical rules and models, must also be developed.

Developing an entirely new set of EDA tools and integrated methodologies, capable of full co-design and multi-physics analyses, is prohibitive. Sooner or later the industry may evolve a separate set of tools specialized for 2.5D and/or 3D co-design and co-analyses; but this will require a lot of innovative development, and an end market with many SiP design starts, with a demand from many customers. In the meantime, most users must expect to leverage their existing investments in (2D SoC) EDA infrastructure, including PDKs, tech files, sign off methodologies and tools, etc., and to extend these via manual or quasi-manual band aids, as required to cobble together a specific SiP product.

4.4.2   Design Methodology Requirements—Practical

In general, the following basic axioms apply to a design flow (2D as well as 2.5D/3D), and are illustrated in the chart in Fig. 4.9

- "Design Eco-System" is a structured methodology, encompassing EDA tools and standardized exchange formats, for moving a design through successive levels of abstraction, down to a polygon level description used for mask making
- Target technology characteristics, in form of rules or models, percolate up the design flow, becoming less accurate, and including more margin, at each level of abstraction
- Target design specification, in a form of a data base and constraints, trickle down the design flow, becoming more accurate and precise after each step in the flow
- The accuracy of all physical models or rules decreases at successively higher levels in the flow

Fig. 4.9   Concept high level view of an IC design flow—a concept flow of a digital design flow, showing the flow of process-related constraints and design-based specifications and highlighting the accuracy and abstraction tradeoffs

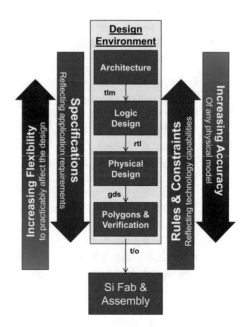

- The ability to affect design change decreases at successively lower levels of the flow

As outlined in Chaps. 2 and 3, it is possible to implement shortcuts in design of some types of 2.5D/3D SiPs, and design each die one at a time, and each of the packaging layers independently of each other, to a fixed spec, and manually hand off various constraints, etc. Possible—but not necessarily desirable, since it would result in excess margin, and inefficiency in the design process—which could very well dilute the value proposition.

Hence, a graceful evolutionary path for development of an MtM design environment that is 'good enough' for the short term, and evolves toward the ideal requirements over time, is required. Within the foreseeable future, design environments capable of meeting all the ideal requirements are not realistic. Too disruptive and too big an effort. A possible practical strategy for shaping the Design Environment for 2.5D/3D Design has been proposed in the past (Radojcic 2008; Milojevic et al. 2009; Radojcic 2012a, b) and is outlined below, for completeness sake. The intent is to enable a graceful evolution of the design methodologies and tools over time. The following are the basic tenets of the proposed strategy:

i. *Manage Choices*: a 2.5D/3D-aware methodology must be able to explore the many degrees of freedom, tradeoffs and 'knobs', outlined in Chaps. 2 and 3, that are offered by these MtM technology options. Experience with regular 2.D SoC designs is that most of the design re-do's, that drive up the cost of design, are due to instability in product specifications. With the range of options offered by the new integration technologies, 2.5D/3D SiPs are especially prone to this. Product specifications must be defined and firmed up, and optimized for a given application in a given technology. It is believed that finding that 'sweet spot' requires multiple iterations of a trial 2.5D/3D designs. The methodology that enables easy trail design is called here *PathFinding*, with the following characteristics:

- PathFinding needs to understand the overall 2.5D/3D architecture, and interact with a proposed design in terms of specs. As such, PathFinding does not need a lot of detailed inputs.
- PathFinding needs to output comparative information about relative 'goodness' of design x versus design y, and hence consistent and relative fidelity versus absolute accuracy, is required. Since PathFinding is done early in the technology life cycle, ahead of actual product design, this bypasses the need for mature and accurate technology models.
- PathFinding is intended to be an iterative process and hence it favors ease of use and design turn-around time over accuracy.
- PathFinding needs to be globally and physically aware and must include models for the 2.5D/3D technology features (TSV, front vs backside, uBump…), i.e., cannot be done purely in a spreadsheet.

- PathFinding does not need to be detailed so frozen IP blocks, chip utilities such as PDN or DFT, etc., can all be accounted for in terms of placement or congestion blockages, without having to be designed and laid out.
- The output of PathFinding is a specification—especially of the kind of design and architecture 'knobs' associated with the 2.5D/3D technologies described in Chaps. 2 and 3. This specification is then an input for Design Authoring.

Performing trial designs in standard 2D tools is also of course possible, and is, in fact, often done. However, these tools are optimized for accuracy and require detailed PDKs to fuel them, so that even without introducing the necessary 2.5D/3D features (multiple die and PDKs, 2 sided die, TSV's, etc.), their use in pathfinding is cumbersome making iterative exploration quite lengthy. Thus, in reality, a specialized tool is required to meet the objectives for PathFinding.

ii. *Manage Interactions*: the effect of various interactions—thermal, electrical or mechanical—precipitated by the 2.5D/3D technologies must be explored, constrained and designed out of SiP products right from the start. Since, right now, it is too hard to implement mesh based multi-physics solvers in EDA design tools, an offline, physically aware multi-physics methodology is required to explore the interactions and to output physical design rules or models that can then be used in a normal design environment. Experience with regular 2D SoC design is that most of design re-spins, that also drive up the cost of design, are caused by instability in PDK and process design rules. With the range of options offered by the new integration technologies, 2.5D/3D SiPs are especially prone to this. Hence a methodology that enables these analyses is required, and is called here *TechTuning*, with the following characteristics:

- TechTuning does not have to be a single tool, i.e., each of the interaction types (electrical, thermal, mechanical) can be assessed in a separate specialized domain. The challenge of doing full multi-physics analyses is unmanageable
- TechTuning can be based on mesh-based tools that model thermal flows (CFD), mechanical stresses (FEM), and electrical fields. These computationally intensive methodologies can be implemented offline, outside the EDA design environment
- TechTuning analyses are performed on suitable 2.5D/3D test structures and Figure of Merit (FoM) circuits that are representative of the die, packages, circuits (die size, power density, etc.) expected to be found in the actual SiP product
- The output of TechTuning is Placement and Layout KoZ rules and/or compact models—which can be uploaded into regular 2D Design Authoring tools (floorplanning, routing, layout, SPICE…) and the associated PDKs
  Note that development of compact models to describe the multi-physics interactions is also possible—and ultimately the right way of importing these considerations into Design Authoring flow. Use of lumped SPICE models

described in Chap. 3 is an example. Compact models capable of analyzing a complete chip level GDS for eCPI stress interactions has been demonstrated in Mentor Caliber tool (Sukharev et al. 2016). Use of abbreviated thermal modeling approach based on design files has been demonstrated by for example, Gradient (now Synopsys). Use of such physically aware compact model methodologies inside a design flow, rather than rules based approach described above, is a better way of importing new constraints. If and when these methodologies are adopted, 2.5D/3D SiP design optimization without a lot of excess margin will be easier than the TechTuning flow.

iii. ***Manage Costs***: in order to extend the existing investment in the current 2D design environment and tools, 2.5D/3D technologies must be described through a set of rules and models, and MtM SiP products must be implemented via a series of 'one-die-at-a-time' 2D chips and package specs + interface specs. This reducing of 2.5D/3D paradigm to 2D framework defined by the existing EDA tools is called here *Design Authoring*, with the following characteristics:

- Design Authoring tools and flows may need to be modified in order to be more useful for design of 3D stacks—mostly in the analyses portion of the flow. That is, whereas the physical design of 3D stacks can be reduced to a set of 2D designs, the final signoff and analyses of the performance of the product must comprehend the entire stack. For example, PDN analyses should be conducted for the entire 3D stack, rather than assuming that the PDN performance on each die is independent of the other die.
- In some cases, even the requirement for the analyses of the integrated stack may be relaxed—depending on given product specifics. It may be possible to insert manual edits or leverage TechTuning to generate implementation rules with sufficient margin to obviate need for final analyses. For example for low power MoL products, integrated PDN analyses may not be required, as described in Chap. 3.

The intent of the proposed strategy is therefore to produce the detailed chip (and package) designs, down to GDS2 (Gerber) level, one die (layer) at a time, and to do the SiP integration through quasi-manual custom design. The risk of exploding design cost due to re-spins and/or redos is minimized by leveraging the PathFinding flow to address churn in specifications, and the TechTuning flow to address instability in process technology. That is, per this strategy, the idiosyncrasies associated with the MtM technology (2.5D. Hybrid, 3D…) are comprehended in the specialized PathFinding and TechTuning tools only, while the bulk of the actual product design is implemented in an unaltered 2D (3D-unaware) Design Authoring Flow, as illustrated in Fig. 4.10.

This way, the bulk of the EDA cost—which is almost entirely invested in the Design Authoring type of tools—is fully leveraged. At the same time, the risks of spec iterations and/or process technology instability—the principal causes of ballooning design cost and schedules—is minimized. Note that SiP products targeting MtM technologies are especially prone to spec iterations and technology instability,

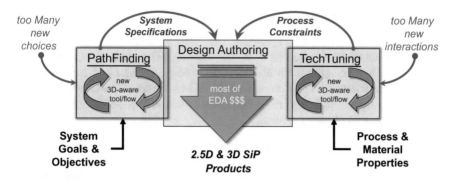

Fig. 4.10 Design-for-MtM strategy—a cartoon that illustrates the proposed concept strategy for design for More-than-Moore 2.5D and 3D SiP, and that defines the role for PathFinding, TechTuning and Design Authoring subflows

partially due to their immaturity, and partially due to number of degrees of freedom and tradeoffs associated with these technology options. With this approach, new EDA tools are required only for the PathFinding and TechTuning flows.

Note that the methodology described here is adequate for some types of 2.5D/3D SiP product designs—as defined in Chap. 3, and roughly corresponding to the current status of the industry. If and when the volume of 2.5D/3D SiP designs justifies it, the EDA industry will deploy a fully automated and integrated co-design flow that explicitly recognize new features and deals with the interactions in situ. However, it is believed that even when this new generation of tools becomes available, they will necessarily first be introduced into the PathFinding flow—due to the relaxed requirements of design planning versus design authoring.

Note also that the interactions between the die in a 2.5D or 3D SiP stack, and especially the multi-physics interactions, can affect the apparent product variability, and hence yield. For example, considering just the thermal interactions in a two die 3D stack, multiple alignments of hot spots can be expected, i.e. there can be Hot-on-Hot, Hot-on-Cold, Cold-on-Cold, and Cold-on-Hot combinations between the hot spots on the two die. Each combination will result in different local temperature. In addition, variability in the thermal conductivity of the SiP package—driven for example by variability in die thickness, or TIM thickness, or substrate Cu line width and thickness, power density, etc.,- can be expected as a part of normal statistical process distribution. Thus, at any given location, a distribution of temperatures, with an intrinsic (design dependent) and extrinsic (process control dependent) component, should be expected—resulting in a distribution of electrical performances. This variability is not comprehended in the standard Process-Voltage-Temperature 'corners' used for signoff of each individual chip. Note that similar interactions can be expected with mechanical stress and electrical coupling, resulting in further broadening out of performance variability. The proposed methodology does not address this increased variability, and perhaps the standard process corners can be margined up to account for this.

4.4.3 Design Methodology—Current Status

A general summary of the status of the various EDA tools for design of the 2.5D/3D SiPs, leveraging the proposed methodology, is given in the tables below. Note that the focus is on implementation of a Heterogenous X-o-L type of design, but with restricted flexibility in T2 floorplan, as this is expected to be the type of design to drive EDA development.

 i. *PathFinding*: key requirement for PathFinding tool is ease of use—in order to quickly implement a trial design and ascertain impact of some input technology or design variable. Currently available EDA tools that are applicable are summarized in the Table 4.4.
 ii. *TechTuning*: The key attribute of TechTuning is to abstract multi-physics challenges into relatively simple layout rules and guidelines, which can be absorbed into normal design tools, by performing the multi-physics modeling and simulations offline, in suitably specialized environment. Currently available EDA tools that are applicable are summarized in the Table 4.5.
    Note that the TechTuning tools and methodologies must be calibrated and validated versus Si, using suitable test vehicles.
 iii. *Design Authoring*: The objective of Design Authoring is obviously to produce a design that will function to spec in Si with minimum die size and complexity, and maximum yield. Currently available EDA tools that are applicable are summarized in the Table 4.6.

Table 4.4 EDA tools for pathfinding—a table summary of the current EDA tools useful for PathFinding trial 2.5D and 3D SiP concepts at the Logic, Layout and Electrical Analyses levels

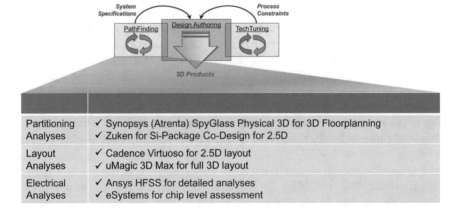

Partitioning Analyses	✓ Synopsys (Atrenta) SpyGlass Physical 3D for 3D Floorplanning ✓ Zuken for Si-Package Co-Design for 2.5D
Layout Analyses	✓ Cadence Virtuoso for 2.5D layout ✓ uMagic 3D Max for full 3D layout
Electrical Analyses	✓ Ansys HFSS for detailed analyses ✓ eSystems for chip level assessment

Table 4.5 EDA tools for TechTuning—a table summary of the current EDA tools useful for TechTuning analyses of 2.5D and 3D SiP constructs for Thermal, Mechanical and Electrical interactions

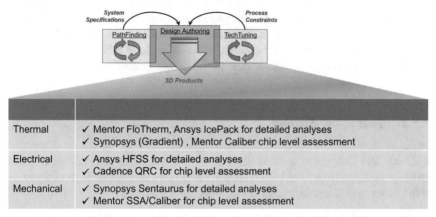

Thermal	✓ Mentor FloTherm, Ansys IcePack for detailed analyses ✓ Synopsys (Gradient) , Mentor Caliber chip level assessment
Electrical	✓ Ansys HFSS for detailed analyses ✓ Cadence QRC for chip level assessment
Mechanical	✓ Synopsys Sentaurus for detailed analyses ✓ Mentor SSA/Caliber for chip level assessment

Table 4.6 EDA tools for design authoring—a table summary of the current EDA tools useful for one-die-at-a-time Design Authoring of digital die for 2.5D and 3D SiP products

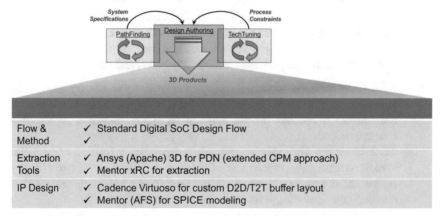

Flow & Method	✓ Standard Digital SoC Design Flow ✓
Extraction Tools	✓ Ansys (Apache) 3D for PDN (extended CPM approach) ✓ Mentor xRC for extraction
IP Design	✓ Cadence Virtuoso for custom D2D/T2T buffer layout ✓ Mentor (AFS) for SPICE modeling

4.5   More-than-Moore Modeling and Characterization Opportunities

In general, test chips are the primary, and often the only, learning vehicles used to develop a technology. For a disruptive technology, in particular—such as More than Moore 2.5D and/or 3D integration technologies—test chip development is especially challenging and critical. "Challenging" because test chips are designed before a technology is characterized, and in the absence of historical background, this calls for some engineering guesses. And "Critical" because the test chip data is a foundation which shapes the overall perception of the value and risks of a given technology. As such, test chips and the associated data are a vital part of the 2.5D and/or 3D eco-system required to deploy MtM technology options. Hence, the test chip development and characterization methodology is briefly discussed here.

4.5.1 Test Chip Strategy

2.5D and 3D integration with its blend of Si and Packaging technologies, should leverage best practices from both domains. Hence an overview of the traditional test chip design and characterization methodologies is pertinent.

i. **Si Technology Test Chip Practices**: More-Moore Si technologies are used to a regular cadence of new technology nodes every couple of years, and the test chips are very expensive—costing several $M's for a full flow vehicle (masks, materials and engineering effort), with ~ 1 year cycle of learning. At that cost, and cycle of learning, test chip redesign is not acceptable. Hence full flow test chips have to be planned for, and are typically structured to extract broad information from each vehicle—often at the expense of efficiency. Typical development methodology uses a blend of test vehicles, each covering a particular portion of technology development life cycle, and intended to derive a given type of information, as illustrated in Fig. 4.11, and outlined below:

- TC-1 Class for Early Module Development: typically short loop test vehicle, with one or few mask layers, compatible with the metrics required to characterize a specific process module (could be physical or electrical metrics, or both…).
- TC-2 Class for Technology Integration: typically, a full flow integration vehicle containing discrete test structures (transistors, transistor arrays, chains and combs, etc.) used to calibrate the PDK and models (e.g., SPICE models) and demonstrate intrinsic yield and reliability.
- TC-3 Class for IP Qualification: typically, full flow, product-like test vehicle containing integrated IP (S/C libraries, SRAMs, DAC's, PHY's, etc.) used to validate all the models (Inc. corner models) and statistically demonstrate yield and reliability.

ii. **Si Packaging Test Chip Practices**: Packaging technologies are used to gradual and evolutionary development and test vehicles tend to be relatively cheap—costing several $10 K's in terms of masks, materials and engineering effort—with a cycle of learning of a few months. Hence the test vehicles are typically

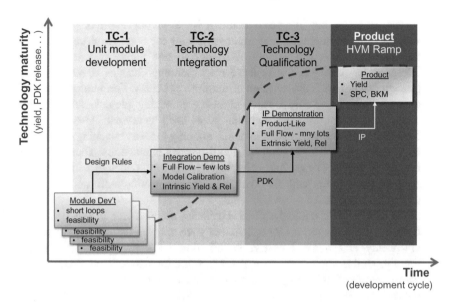

Fig. 4.11 Si technology test chip strategy—a concept chart of technology maturity versus calendar time, identifying the different classes of test chips typically used to develop a Si Technology, and highlighting the key information derived from each type of a vehicle

more ad-hoc and tend to be focused on extracting specific, relatively narrow information, if and when needed. At these costs the test vehicles can be iterated through successive cycles of learning, i.e., redesign of test chip is acceptable and practiced. Typically, the test vehicles break down into tree types:

- mechanical Test Vehicle (mTV) typically using a dummy Si coupon to check the physical integrity of a given new feature of a packaging technology, with physical metrology only
- electrical Test Vehicle (eTV) typically using a Si test chip with one or two metal layers discrete structures to check the electrical characteristics (shorts, opens, R-L-C, chains…) of a package family
- product test vehicle typically using an active product die to demonstrate the overall integrity and yield of a package family, leveraging product test metrology

iii. *More than Moore Test Chip Strategy*: the test vehicles for characterization of 2.5D and especially 3D technologies in a way inherit the worst of both worlds. They are typically too expensive to be the 'narrow' vehicles that could be easily iterated through design → characterization → redesign cycle, but there is no structured practice and precedent for definition of a vehicle for capturing 'wide' data—like a TC2 or a TC3 above. There is no simple set of target rules, or ITRS roadmap, to dictate the requirements. The cost issues are further compounded by the integrated nature of SiP products—involving multiple sources of Si technologies and packages—with each one requiring a suitable, but interdependent, test vehicle. A change in one source (e.g. active die or interposer or package...) may drive corresponding changes at other sources, thereby driving potential redesign of multiple test chips. Furthermore, the PDK used to design a test chip is based on best engineering guesses, but without an opportunity to re-spin the design, the guess must be correct. Also, in practice these test vehicles are intended to define not only the basic stacking design rules and models, but also the complex, multi-physics interactions. Thus, designing test chips for 2.5D and 3D technologies is a big challenge and requires a separate eco-system, with its own practices, engineering skill mix, infrastructure, etc. A big deal

Since with MtM technologies there is neither a roadmap nor an experience base from last node, the targets for a technology are not well defined, and there is no uniformity across the industry. PathFinding takes its place—as described throughout this book. And since Test Chips are intimately intertwined with the methodology of technology development, the test chip strategy needs to change as well. Specifically, in order to optimize 2.5D or 3D integration technologies, via the pathfinding studies, for a given target architecture, Test Chip development should be integrated and coordinated with PathFinding and TechTuning. Hence it is believed that the technology development flow, product architecture development, and the user–supplier handoffs all need to be adjusted for successful deployment of MtM technology options. The iterative pathfinding development environment favors a strategy with the following attributes:

- early and ongoing collaboration between the design and technology development activities
- regular asynchronous updates of technology DRM, Models, PDK...
- regular asynchronous update of technology targets
- fixed schedule for regular and frequent release of test chips
- variable test chip content
- parametrized test chip design that allows quick tuning to revised rule set

That is, a test chip strategy based on regular test chip tape outs, but with variable content that reflects the latest learning, is believed to be the right strategy for 2.5D/3D technologies co-development. This is illustrated in the concept flow shown in Fig. 4.12. Thus, the tradition where TC-1 (module development), TC-2 (technology characterization) and TC-3 (IP qualification) are pushed into the foundry/OSAT realm, and product architecture is abstracted and pushed into a separate IC Design realm is not optimal for 2.5D/3D integration technologies.

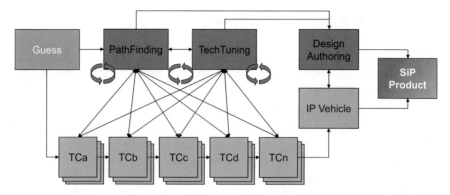

Fig. 4.12   2.5D and 3D integration technology test chip strategy—a concept chart illustrating a proposed strategy for use of test chips for development of 2.5D and 3D integration technologies, and highlighting the coupling of PathFinding and TechTuning design flows with test chips, taped out on a regular cadence

4.5.2   Test Chip Content

The nature and scope of the test vehicles needs to be supplemented in order to capture and describe some of the 2.5D/3D technology characteristics. Specifically:

- Expand the Scope: the scope of the test vehicles—especially TC-2 class—needs to be extended to encompass characterization of the multi-physics interactions. Structures for calibration and validation of thermal (heat sources, thermal sensors) and mechanical (strain sensors) models in addition to the traditional electrical and physical models, are required.
- Expand the Definition: incremental test vehicles focused on a specific mechanism, rather than a specific technology—may be merited. For example, generic, technology independent and pre-designed, test chips for thermal or stress characterization are needed.
- Expand the Metrology: in most cases test chips are designed for electrical measurements—favored due to speed and accuracy of electrical tests. There is a need for test vehicles that rely on other measurement techniques, such as for example TEM, X-ray, cantilever curvature, wafer bending, nanoindenting, etc., all of which require specialized test structures suited to specialized metrology.
- Expand the Statistics: typically test chips are excellent vehicle for accurate characterization of a given attribute, but are lousy vehicles for generation of statistics that sample across all process corners and variables. That is because test chips are a cost and not a revenue item, so that they are run only on a (few) sample lots. Hence reduced test structure set including features relevant to 2.5D and 3D integration—e.g., TSV, uBump, BRDL, D2D fine pitch wiring, etc.—are required and can be sprinkled in the scribe lanes, or within the actual product floorplan.

4.5.3 Characterization and Models

In addition, some of the model requirements—which in turn drive the test chip characterization requirements, also need to evolve. The attributes favored by MtM type of technologies are:

- Use of Guess Models: pathfinding is based on tradeoff analyses, which, requires simulations and models. Waiting for discrete releases of rev 1.0 or even rev 0.1 models to start pathfinding may be too late—especially if pathfinding is to influence the technology development. Ergo 'best guess models' are required—and must be used fearlessly.
- The risks involved in making architectural decisions based on guess models must be managed and mitigated, preferably through frequent updates of models based on latest (frequent) updates of characterization data.
- Use of Parametrized Models: modeling methodology that enable easy change to the appropriate equation constants—representing some attribute a technology—is required, in order to enable the iterative pathfinding studies.

References

Elfadel A, Ibrahim M, Fettweis G (eds) (2016) 3D stacked chips. Springer. ISBN 978-3-319-20481-9

Ho P, Hu CK, Nakamoto M, Ogawa S, Sukharev V, Smith L, Zschech E (eds) (2014) Stress induced phenomena and reliability in 3D microelectronics. In: AIP conference proceedings 1601, May 2012, Kyoto, Japan. ISBN 978-0-7354-1235-4

Milojevic D, Radojcic R, Carpenter R, Marchal P (2009) Pathfinding: a design methodology for fast exploration and optimisation of 3D-stacked integrated circuits. In: IEEE 2009 international symposium on system-on-chip (SoC 2009)

Papanikolaou A, Soudris D, Radojcic R (eds) (2011) Three dimensional system integration: IC stacking process and design. Springer, US. ISBN 978-1-4419-0961-9

Radojcic R (2008) Design infrastructure for 3D TSS architectural exploration & design implementation. http://www.sematech.org/meetings/archives/3d/8518/pres/Radojcic.pdf. Accessed 26 Nov 2016

Radojcic R (2012a) Roadmap for design and EDA infrastructure for 3D products. In: IEEE hot chips HC24, August 2012

Radojcic R (2012b) Roadmap for design and EDA infrastructure for 3D products. In: IEEE electronic design process symposium, April 2012, Monterey, CA

Radojcic R (2014) Managing thermal & mechanical interactions with 2.5D and 3D IC's. In: iMAPS advanced technology workshop and exhibit on thermal management, Los Gatos, California

Topaloglu RO (ed) (2015) More than Moore technologies for next generation computer design. Springer, New York. ISBN 978-1-4939-2162-1

Sukharev V et al (2010) 3D IC TSV-based technology: stress assessment for chip performance. AIP Conf, Proc 1300

Sukharev V et al (2016) CPI stress induced carrier mobility shift in advanced silicon nodes. In: Proceedings of the ASME 2016 international mechanical engineering congress and exposition (IMECE2016), November 2016, Phoenix, Arizona, USA

Tengfei J et al (2015) Through-silicon via stress characteristics and reliability impact on 3D integrated circuits. MRS Bull 40

Wu B, Kumar A, Ramaswami S (eds) (2011) 3D IC stacking technology. McGraw-Hill, US. ISBN 007174195X

Xie Y, Cong J, Sapatnekar S (eds) (2011) Three-dimensional integrated circuit design. Springer, US. ISBN 978-1-4419-0783-7

Xu X, Karmarkar A (2011) 3D TCAD modeling for stress management in through silicon via (TSV) stacks. In: AIP Conference Proceedings 1378

Zschech E, Radojcic R, Sukharev V, Smith L (eds) (2011) Stress management for 3D ICS using through silicon vias. In: International workshop on stress management for 3D ICs using through silicon vias. American Institute of Physics. ISBN 10: 0735409382

Chapter 5
More-than-Moore Adoption Landscape

The practices used to deploy advanced 2D SoC IC products have been developed over the decades, and the industry is quite familiar with dealing with the risks and challenges of new products that leverage new CMOS technology nodes, as according to the traditional Moore's Law. Practices that are more or less standard across the industry have been developed to manage any technical or sourcing risks, and the characteristics of the future CMOS technology nodes can be predicted reasonably well. On the other hand, the eco-system required to develop and deploy SiP IC products, which leverage the More-than-Moore technology options, has not been proven through to high volume manufacturing, so that SiP implementation for high volume applications is generally perceived as disruptive, and with unknown technical or sourcing risks. The trade-offs associated with the implementation of a disruptive technology option—such as the 2.5D and 3D technology—versus using the incumbent solution—i.e., 2D SoC—are complex, multidimensional and evolving, and vary with industry sectors, product types, business cycles, technology maturity, etc. Thus, in order to understand the factors that gate deploying 2.5D or 3D technology in mainstream IC products it is important to understand the "landscape" that shapes these decisions. The intent of this section is to describe the constraints and the various tools and mechanisms typically used to make the technology trade-offs for high volume products. This "landscape" is presented in three broad categories, addressing the business structure, the technology, and the product considerations, respectively. Note that the focus is on IC products targeting mobile marketspace—which is currently the largest single market, as illustrated in Fig. 5.1 below for 2014.

The chapter is focused on the process of making decision to adopt a "disruptive" technology versus an "incumbent" technology. Clearly, there are shades of gray, and no technology is entirely disruptive or entirely evolutionary, but these are defined here as the following:

© Springer International Publishing AG 2017
R. Radojcic, *More-than-Moore 2.5D and 3D SiP Integration*,
DOI 10.1007/978-3-319-52548-8_5

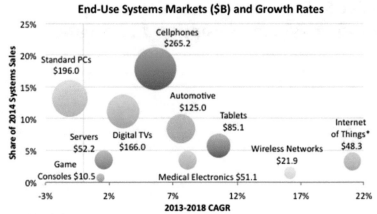

Fig. 5.1 Semiconductor Market Share and Growth Rates—a chart showing (2014) share of the total IC revenues versus the predicated Compound Annual Growth Rates for various application sectors. As shown mobile market is currently the largest single sector

- **Incumbent Technology**: technologies that are an evolutionary derivatives of an existing mainstream technology used in current products in an existing market, and with an easy projected value proposition. For example, scaling CMOS technology from node n to node n+1, along the More-Moore path, is an incumbent technology solution. Similarly, pushing package substrate technology from 10 um to 8 um line/space spec purely through improvements in process and materials is an incumbent technology solution. And so on.
- **Disruptive Technology**: technologies that break a trend in technology evolution and/or roadmaps and require a change in product architecture and/or the design to generate a value proposition. For example, TSV-based 3D stacking is a disruptive technology solution, WideIO memory is disruptive relative to LPDDR memory, and so on

5.1   Business Development Landscape

Electronic Systems industry is comprised of a variety of company types with a given set of core competencies, pursuing a variety of different business models. The entire 'food chain' involved in realizing a system—e.g., a Smart Phone—is very complex, with multiple relationships (suppliers, buyers, partners, etc.), different skills (e.g., Software vs Hardware), different elements in the supply chain (e.g., EDA, foundry, OSAT, etc.), different sourcing models (e.g., parts, boxes, services, etc.) and so on. Thus, for example, development of just the hardware side of a given smart phone involves mechanical (the case, the display, the overall form factor, mechanical and thermal behavior, etc.), electrical (PCB, battery, all components, antennas, etc.), and Si engineering (all ICs such as memories, processors, PMICs, etc.). Si components may be commodity IC products (e.g., memories) sourced by one type of company, versus custom or ASIC/ASSP products sourced by different type of a semiconductor company. Those companies in turn may source their ICs from their internal fabs, packaging and test facilities, or use external foundries and/or OSATs. The fabs—internal or commercial foundries—may acquire masks, equipment, and chemicals from another tier of specialized entities, and so on. The food chain of engineering skills, products, and services is big and long, and a company may choose to compete in one, some, or maybe even most of the disciplines. The nature of the business structure and the associated business model drives how a given company may weigh the risks and benefits associated with the implementation of a disruptive technology option versus incumbent technology options.

5.1.1 "The Cast"

For the purposes of understanding the approach to possible adoption of 2.5D or 3D SiP technology options, some broad-brush definitions of company types, with a focus on Si and packaging technologies, are described below:

i. **Vertically Integrated System Companies**: those that produce and/or market the end electronic appliance. In the case of phones, for example, this would be companies such as Apple, Samsung, HTC, Lenovo, etc. The focus here is on the hardware design, even though nowadays, it is the software that is a major differentiator in the marketplace. There are several types of system companies:

   - Vertically Integrated Companies with Si Fabs (e.g., until recently IBM)
   - Conglomerates with Si Fabs (e.g., Samsung)
   - Fabless Vertical Entity (e.g., Apple).

   Vertically Integrated Entities with internal fabs are typically best suited for development of disruptive Si and Packaging technology options—because they have an opportunity to realize some system-level value proposition by suitably trading off various technology attributes at lower level of the value chain. In addition, they have the internal engineering skill mix required to make and manage the technical trade-offs. Thus, vertical entities have an opportunity to differentiate their systems by deploying unique or disruptive technology options, and presumably to monetize the option through enhanced system sales.

ii. **Specialized System Companies**: Most of the system entities—examples in the case of mobile phones would include Xiomi, HTC, LG—are not vertically integrated, and tend to leverage external Semiconductor Companies for design and sourcing of the Si components. Note that there is a whole spectrum of nonintegrated system entities, and some outsource not only the Si engineering, but also some or even most of the system engineering to various specialized OEM (Original Equipment Manufacturer) or ODM (Original Design Manufacturer) entities. Thus, Specialized System Companies typically would not have the skill mix to weigh the SiP versus SoC technology trade-off, and rely on their semiconductor suppliers to make this type of decision. They therefore relate to disruptive technology options through specifications, i.e., some chip is cheaper or faster or smaller than its equivalent competitor.

iii. *Semiconductor Companies*: There are also several types of companies competing in semiconductor market—each with its own strengths and weaknesses. These are listed, in order to relate the effect of the basic business structure on the disruptive technology development activity:

- Integrated Device Manufacturers (IDM): design and sell IC products and have in-house fabs, assembly and test facilities, and so on (e.g., Intel)
- Fabless Manufacturers: design and sell IC products, but subcontract manufacturing to foundries and OSATS (e.g., Marvell)
- Fab-Lite Manufacturers: a hybrid of IDM and Fabless models, who design and sell IC products and do both, internal fabrication for some products, and subcontract manufacturing for other products (e.g., Freescale)
- Integrated Fabless Manufacturers (IFM): hybrid of IDM and Fabless model who design and sell IC products using only subcontracted manufacturing, but have in-house technical skills (and scale) to customize some technologies (e.g., Xilinx, Qualcomm)

Since IDM's have in house fabs and packaging facilities, they can most easily pursue differentiated technology options—although these then have to be monetize-able and differentiated at the IC level (not system level). IDM's are typically attracted to disruptive technology options—since it gives them an opportunity to differentiate their products in a way which is hard for their competition to emulate.

Fabless entities tend to differentiate their products purely through design and are averse to differentiated technologies, as they tend to use off-the-shelf technologies offered by their supply chain partners. The strength of the fabless model vis-à-vis leveraging a technology is the diversity of choices offered by the supply chain. They can pick and choose the best in class technology, whereas the IDM's are constrained to use whatever technology they have developed in house.

Fab-Lite and IFM's are in-between, and sometimes do embrace a differentiated technology option. Typically, fab-lite entities started as IDM's and IFM's started as Fabless—and their typical disposition toward technology differentiation is reflective of their corporate DNA, respectively.

iv. *Foundries*: make money primarily by converting CapEx investment and Si technology engineering into cash, through manufacturing and selling wafers (e.g., tsmc, GF, SMIC, etc.). Investment into developing a differentiating technology option is managed to increase wafers price, or wafer volume, or position versus competition. By definition, foundries want to sell wafers to multiple customers, and hence differentiated technologies are a bit of a risk. On one hand, disruptive technology option may command a premium wafer price, or lock in a market share. On the other hand, fabless clients usually prefer to multi-source, and hence want to commoditize a technology as soon as possible. So foundries typically need a committed partner-client to drive development of disruptive and differentiating technology options.

v. **Outsourced Semiconductor Assembly and Test (OSAT)**: make money primarily by converting CapEx investment and Package technology engineering into cash through selling assembly services (e.g., Amkor, ASE, SPIL, etc.). OSATS are structured and motivated much like the foundries, but typically operate at lower margins and hence have less latitude for time-to-money. Furthermore, the variety of technology options is unbounded, and oftentimes much of the engineering is done by the user, thereby diminishing opportunities for OSAT differentiation. Hence OSATS need to believe that a differentiating and disruptive technology option—especially ones that require new CapEx—is in great demand in order to invest in its development. Anything that is likely to result in time-to-money beyond ~ 2 year horizon is probably too risky

vi. **Equipment and Material Vendors**: make money by selling specialized materials or equipment to foundries and/or OSATs (e.g., Applied Materials, Rudolph, TEL, Hitachi, Sumitomo, etc.). Truly disruptive technologies typically require either a new piece of equipment or a new material—or a bit of both. The time to money for equipment vendors, i.e., time to repeat orders of a given piece of equipment, is probably 5+ years (time to develop new equipment + time to sell it + time for a fab to integrate it in a target technology + time to qualify and ramp to HVM). Given the limitations of foundries and OSATS, equipment makers are more likely to be led toward disruptive technologies by IDM type of customer. These entities are not able to integrate a given disruptive technology option into a complete process flow, so that equipment and material vendors—absolutely key elements for any disruptive technology option—necessarily need to partner with someone. Note that, with the long time to money equipment vendors need to have an industry roadmap, as a form of assurance of future demand. ITRS was an excellent tool for driving the equipment vendors. Unfortunately, there is no ITRS-like roadmap for MtM class of technologies (yet).

vii. **EDA Vendors**: (Cadence, Mentor, Synopsis, etc.) make money by licensing access—often measured in terms of number of copies ("seats")—to specialized CAD tools for IC design, and as such are also an essential part pf any disruptive technology option. EDA time horizon is ~ 1–2 years, i.e., it is unlikely that the EDA vendors will invest on their own in a major development of tools that cannot be sold for a few years. This is especially so for large established EDA vendors—who need to feed a $\sim \$B$ business. Similarly, if a disruptive technology requires a small number of seats, then the EDA vendors are unmotivated to invest the development effort. Traditionally, disruptive options were attractive to EDA start-ups, hoping to grow with the technology.

In either case, this sector too needs a driving partner to pull the necessary engineering development effort to enable disruptive technology options

viii. *Academia*: (e.g., Georgia Tech, UCSD, Stanford, MIT, etc.) is one sector that has a long time horizon—certainly longer than ~ 5 years. As such, given that the runway for a disruptive technology option can be 10 to 15 years, academia may be the right entity to pursue development and characterization of disruptive MtM technology options. However, with some exceptions, academia has been priced out of much of the technology development arena, and whereas they have the time horizon, and the talent, they may not have access to the right labs and equipment to produce experimental data. Some of the universities are enabled with well-equipped labs, but they typically focus on revolutionary opportunities (e.g., MIT Labs) and/or have difficulties maintaining consistency required for industry participation

xi. *Consortia*: (IMEC, Sematech, LETI, IME, Fraunhofer, etc.) are specialized R&D entities, usually founded on an academic center, and often at least partially funded by a government, where resources are pooled, and a critical mass required to enable a pilot line, and permanent staff, is realized. Like Academia, the Consortia are focused on pre-competitive research, and as such tend to have a long time horizon to drive disruptive technology development. Unlike Academia, Consortia need and solicit (paying) participation from the industry and tend to have a tight policy on managing the IP generated through the collaborative work. Industry participation, and contribution, also ensures that the focus of the R&D effort is relevant.

5.1.2 "The Plot"

It is becoming increasingly apparent that the technical and economic challenges faced by the traditional More-Moore type of CMOS scaling are becoming increasingly difficult and expensive. Yet, the ever-increasing system level integration, that the market has become used to, will likely continue. However, adoption of More-than-Moore integration class of technologies in commercial products is a disruptive change that poses considerable business risks. So, the business scenario where the MtM technology intersects a product is as important as the technical trade-offs. The intent of this section is to outline the business scenarios when adopting a disruptive technology becomes a favorable proposition. Judging from the past trends, and the years of experience, it is believed that a disruptive technology option can displace an incumbent technology solution under one of the following three circumstances:

i. *Incumbent Technology Runs out of Gas*: the easiest path for product implementation of a disruptive technology is a situation when meeting target specifications become prohibitively difficult by extending incumbent technology solution. When the incumbent solution hits a wall, the risks and costs associated with implementing a disruptive technology option oftentimes become palatable.

Pertinent examples of this circumstance include the use of 2.5D integration with Si Interposer solutions for high end FPGAs, or GPU products, such as practiced by Xilinx and AMD. As discussed in Chaps. 2 and 3, the concerns with feasibility, scalability, and/or cost of achieving the target FPGA density using a traditional approach with a giant mono-die, must have made the challenges associated with Si interposer seem relatively attractive, in case of Xilinx. Similarly, the power efficiency and bandwidth required by the high-end graphics processors must have made HBM 3D stacked memory and a Si Interposer more attractive than the use of traditional GDDRx off-chip memory solutions, in case of AMD. In both these cases, the incumbent technology is running out of gas, and the cost-performance constraints associated with high end market where these products compete created an opening for a disruptive technology.

This type of circumstance is the most common condition for adoption of a disruptive technology option, and typically it is adopted first by the high-end products, which have the tightest performance requirements and most generous margin headroom. And of course, once adapted by the high-end pioneers, the learning generated leads to the maturing of the cost structure of the disruptive technology solution, which oftentimes then trickles down into the mainstream. This was, for example, the case with C4 flip chip bumps, and many other technologies that were a disruption to the then status-quo.

ii. *System Level Value Proposition*: An alternative circumstance is a situation where an incumbent technology can still meet the target specs, but a disruptive technology option offers an opportunity for a better value proposition at the system level. An example of this circumstance is the opportunity offered by the WideIO DRAM stacked on a baseband logic die using TSV technology (the disruptive option) versus using LPDDRx DRAM with a POP technology (the incumbent solution). The power efficiency and form factor advantages offered by the WideIO DRAM technology were very well publicized, and seemed attractive and well positioned especially for mobile applications. Multiple companies competing in this space—including TI, ST, Samsung, and Qualcomm—all invested engineering effort to evaluate this option. However, as discussed in Chap. 3, WideIO with 3D TSV stacking has not been adopted.

Clearly, the advantages of this disruptive technology option did not outweigh the disadvantages. The 3D WideIO MoL SiP however involves trade-offs between increased costs and risks at the *component level* versus. benefits at the *system level*.

- Component Level Cost: the TSV technology required for 3D stacking, and SoC design changes, resulted in a higher cost Application Processor component than the traditional 2D solution. Similarly, WideIO DRAM—especially for densities above 1 GB—costs more than the equivalent traditional LPDDR DRAM. In addition, 3D WideIO stacking involves perturbation to the well-established POP sourcing model, which increases business risks.
- System Level Benefit: The form factor (in x,y, and z dimensions) and power (use-case-dependent) advantages at the component level—universally

perceived as goodness in the mobile sector—can be monetized at the system level, through thinner or smaller phones with better displays and/or battery life, and so on.

Realizing this type of a cross-domain trade-off requires coordination capabilities necessary for full and complete optimization across the supply chain. This can potentially be exploited by a Vertically Integrated entity that has an internal fab (e.g. Samsung). The coordination across corporate boundaries and integration of the distributed Supply Chain, involving OEMs (LG, Xiomi, etc.), semiconductor entities (Qualcomm, Mediatek, etc.), DRAM vendors (Micron, Hynix, etc.), and so on, makes it difficult to implement for fabless, nonintegrated, system entities. Clearly, in this specific market, a scenario where cost (and risk) is on the side of the incumbent technology solution trumps all other advantages offered by a disruptive technology candidate. That is, when the incumbent solution can meet the requirements, cost is the dominant system level value proposition—at least in mobile space.

iii. ***Component Level Value Proposition***: The third circumstance for intercepting products with a disruptive technology option is a situation where an incumbent technology can meet the target specs, but a disruptive technology option offers an opportunity for better cost at the component level.

An example of this circumstance is use of a multi-chip packages, either to bundle existing components in a single package (integration value proposition in Chap. 2), or to enable breaking up of a complex SoC into several smaller die (Split Die value proposition in Chap. 2). However, as discussed throughout this book, finding that "sweet spot" requires analyses and tradeoffs. Achieving lasting value proposition with a disruptive option may be beneficial to cost—but involves risks. Managing risks is a complex business exercise, and taking risk with a company's mainstream product line is not a desirable strategy. Therefore, even if cost is on the side of the disruptive technology option, it does not necessarily trump the risk advantages offered by the incumbent technology solution—at least in existing mainstream product lines.

Thus, the business scenarios that favor use of disruptive 2.5D and 3D technology SiPs are summarized in Table 5.1.

Table 5.1 2.5D and 3D Technology Intersect Scenarios—a table summary of the types of business scenarios that favor use of disruptive 2.5D and 3D technology SiPs versus the incumbent 2D SoCs, highlighting the nature of the business and the differentiating attributes that are expected to drive the adoption

1. Incumbent Runs Out of Gas	2. Opportunity for Better System	3. Opportunity for Better Component
• Pro: IC Performance and scalability	• Pro: System Attributes, e.g., power, form factor	• Pro: IC Attributes, inc., cost-performance
• Con: Cost	• Con: Monetizing it	• Con: Risk
• Open to High End Applications: Attracted to performance and have margin to increased cost	• Open to Vertical System Entities: Attracted to differentiation (=monetization) of system	• Open to Chip Vendor Entities: Attracted to lower cost structure

5.2 Technology Development Landscape

5.2.1 Landscape for "More-Moore" Scaling

Under most conceivable scenarios for the More-More path, radically different materials, equipment, and device architectures will be required, and hence the costs and R&D horizon, can be expected to continue rising. Time horizon of ~ 5 to 10 years for pre-competitive R&D + $\sim \$2B$ per node for engineering development + $\sim \$10B$ for a new fab. With the rising costs of manufacturing and R&D required to stay on the MM curve, semiconductor industry has evolved toward

(a) shared costs of manufacturing, either via the fabless-foundry model resulting in companies sharing the cost of foundries, or through continued consolidation resulting in more products filling in-house fabs, or a mix of these, and

(b) shared pre-competitive R&D, via research consortia (e.g., IMEC, ITRI, LETI, etc.).

Note that the continuing consolidation in the industry is not indicative of backing away from the concept of shared pre-competitive R&D costs. Thus, for example, Intel and Samsung—the dominant IDM's, Qualcomm, and Xilinx—the dominant fabless players, etc., are all very active members of the Semiconductor Consortia such as IMEC. That is, it is believed that More-Moore paradigm will continue to be sustained through collaborative pre-competitive R&D via research consortia on the development side, and more consolidation on the product and manufacturing side.

5.2.2 Landscape for "More-than-Moore" Scaling

If the traditional More-Moore scaling is bottoming out, then the real question is how to adapt to the next paradigm. An Observations quoted from 2009/2010 ITRS is pertinent.

"Since the early 70's, the semiconductor industry ability to follow Moore's law has been the engine of a virtuous cycle: through transistor scaling, one obtains a better performance–to- cost ratio of products, which induces an exponential growth of the semiconductor market. This in turn allows further investments in semiconductor technologies which will fuel further scaling. The industry is now faced with the increasing importance of a new trend, "More-than-Moore" (MtM), where added value to devices is provided by incorporating functionalities that do not necessarily scale according to "Moore's Law"

More-than-Moore class of technologies—things like heterogeneous integration, 2.5D and 3D SiPs, etc., are likely the next paradigm. Since the value proposition of the More-Moore scaling for the digital sub-system is continuing, More-than-Moore class of technologies must be co-developed and integrated with the scaled CMOS technologies, that is, it is not like MM R&D will stop.

Hence, up to now the development of the MtM class of technologies has followed two separate paths: namely, through the Si domain, for integration of new devices and/or features in baseline CMOS SoC, and through the Packaging domain for integration of different die in a SiP package.

i. *Si Domain* (IDM's and Si foundries): the obvious example of the MtM technologies in the Si domain are development of Through-Si-Via, or the development of Magnetic Tunnel Junction (MTJ) devices (in pursuit of embedded non volatile memory opportunity). However, experience to date indicates that these efforts tend to take a back seat w.r.t. the (growing) effort required for the traditional MM scaling. To date, the engineering effort and the development schedule required to develop a leading edge CMOS node + an MtM technology feature + a demonstration of the integration of these features, seems to be prohibitive. Hence, it is believed that there are no efforts anywhere in the industry to deploy a TSV module in any leading-edge CMOS node beyond 14 nm. Currently, if anything, the development efforts for MtM technology options are typically pursued as midlife kicker option for an established CMOS node.

   However, the cadence of mid-life kicker technology option is at odds with the cadence of new product introduction—and consequently there has been no intersection of MtM options with mainstream digital products. That is, new product architectures are typically introduced to coincide with the release of new CMOS nodes. By the time a technology mid-life kicker becomes available, the architecture is pretty much "closed hood"—and product tape outs are typically trickle-down implementations of existing architectures. Business (cost, time to market, market share, etc.) and Technical (power-performance) considerations —at least in the mobile market—indicate that re-architecting a product to leverage a mid-life kicker technology option is not as good as re-architecting the product to intersect a next node.

   Consequently, leading edge TSV development is abandoned or suspended in all major foundries. There are no plans for MtM options in 10 nm or 7 nm nodes anywhere (to the best of author's knowledge). The one exception is 3D DRAM memories—as discussed in Chap. 3.

   That is, it is believed that the MtM technologies for heterogenous integration sourced from the SI Domain will first be offered on older nodes targeting components that do not push the CMOS technology—such as RF, analog designs, integration of sensors, etc.

ii. *Packaging Domain* (Substrate and OSAT Service suppliers): development of MtM technology options in assembly and packaging domain is an obvious way of addressing the engineering resource conflicts encountered in the Si domain, i.e., it is a way of separating the resource investment required for More-Moore scaling from the resource investment required for More-than-Moore scaling. Consequently, there is more traction in the industry for adoption of package technology based SiP avenues for implementation of the MtM technologies. The OSAT industry has in fact embraced 2.5D technologies and all major providers have made significant investments—both, in terms of CapEx and the Engineering effort—to enable various flavors of 2.5D or 3D SiP technology options. However, without minimizing the challenges or the accomplishments, the key enabling MtM technologies currently deployed by the OSATS, such as 40 u pitch uBump attach, 2.5D Si interposer, Fan Out technology, 3D stacking (w/o TSV), embedded substrate, etc., are all low hanging fruit developed by extension of the current 2D state of the art.

That is, it is likely that the MtM technology options sourced from the Packaging Domain will be evolutionary implementations, rather that some disruptive revolutionary option. This is exacerbated by the need a design eco-system, since OSATs traditionally are not tightly coupled to the EDA sector.

5.3 Product Development Landscape

Technology adoption decisions made by a product manager are, understandably, shaped by business and market considerations as much as by technical factors. The potential Value Propositions normally associated with adoption of a candidate technology (power, performance, form factor) has to be balanced versus the risks, cost, and schedule challenges associated with the given technology option. These trade-offs are obviously shaped by the position of a given product in its market, i.e., a product that is an established market leader in a given segment is likely going to make a different decision than a product that is just breaking into a (new) market segment. Furthermore, the decisions are also affected by the market itself and the associated business cycle—growing markets typically favor a different trade-off than shrinking markets—and tend to strike a different balance between the Benefit versus Risk and Cost sides of the equation, and so on.

This section outlines the parameters that are typically explored when making a technology decision for a given product. Note that the emphases is, again, on products competing in high volume consumer market—such as mobile.

5.3.1 Technology Intersect

In general, at a high level of abstraction, an IC Product Development Cycle, could be viewed as a four-phase process, as following:

• Product Definition—initiation phase where the target market driven requirements are defined, typically resulting in some form of a Market Requirement Document (MRD). This phase is owned by the Business Development and

Marketing teams, and sets the project cost and schedule objectives and defines some of the high level functionality, performance, and form factor constraints

- Product Architecture—initial design phase where the top-level architecture is defined, typically resulting in some kind of an Architecture Spec. This phase is owned by engineering teams and defines the chip functionality, power-performance, IP content, interface standards, form factor, etc., and selects the target Si technology and package for a given SoC.
- IC Design–full design phase where the architecture specification is actually implemented in a SoC design in the chosen Si technology/Package, and resulting in a tape out (T/O). This is clearly an engineering function
- IC Ramp–final phase that includes design verification and Si debug, test optimization, supply chain management, yield optimization, etc., that results in an IC product that, presumably, meets the market requirements. This phase is typically where the engineering team hands off the product to the operations team.

Typically, a new SoC design is authored for every CMOS technology node, and depending on the nature of the target technology, there is some tolerance in *when* a specific technology decision intersects product design cycle. For example, More-Moore type of Si technology must be selected—at the latest—at the *start* point of the product design phase. Note also that typically, evolutionary technology decisions are made *as late as possible*, as this minimizes potential risks (and costs).

In case of a disruptive technology—such as More-than-Moore class of technologies—the candidate technology selection must be made at the beginning of the "architecture" cycle, to allow time for pathfinding, trade-off analyses, and architecture and technology co-optimization. This is because, as discussed throughout this book, *the full value proposition of a disruptive technology can be realized if, and only if, the target product is architected for it.* That is, in practice, the selection of a disruptive More-than-Moore technology has to be made earlier in Product Development Cycle than the corresponding decisions with the evolutionary More-Moore technologies.

This concept IC Product Development Cycle is illustrated in Fig. 5.2.

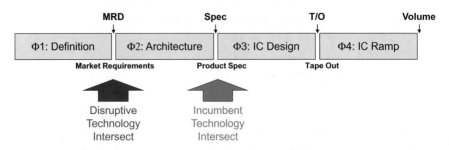

Fig. 5.2 IC Product Development Cycle—a concept high level chart showing the phases of an IC Product Development Cycle typically practiced to source high end mobile SoC products, from concept to volume manufacturing, and highlighting the "intersect point" for fundamental technology decisions

5.3.2   Schedule Conundrum

Typical (optimistic) durations required for the different phases of the Product Development Cycle for a modern leading edge SoC product, such as used in the mobile phones are (in reverse order from HVM):

- Ramp (Tape Out -> Silicon Prototype and Yield Ramp -> Customer Shipment) ~ 12 months
- Product Design (Spec -> Logic Design -> Physical Design -> Tape Out) ~ 12 months
- Architecture (Market Requirement -> Spec) ~ 12 months

That is, a SoC product commitment to a disruptive technology option needs to be around ~ 3 years before it is ramped into volume manufacturing.

The development runway, even for a relatively simple SiP–class of packaging technologies (e.g., advanced substrates, interposers, etc.) is of the order of 18 to 24 months—from concept to initial demonstration. Development time for disruptive technologies based on a Si baseline technology (e.g., TSV, MRAM) is longer. That is, it takes a few years for a technology to get through module level assessment, and move to full qualification, especially if it is not pushed by a compelling product need. Therefore, in practice, in order to allow development time, work on a candidate disruptive technology option needs to be initiated 2–3 years ahead of the product intersect, adding up to a total of 5–6 years ahead of the volume ramp.

However, a high volume commercial product typically hesitates to commit to a disruptive and new technology ahead of a demonstration, preferably including yield and cost data generated during volume manufacturing. Similarly, in the absence of a product commitment, supply chain entities are reluctant to spend the CapEx and acquire the equipment required by a disruptive technology, that far ahead of opportunity to get an ROI. This is the chicken-and-egg conundrum mentioned in Chap. 4.

Thus, the combination of a long technology development runway and risk aversion of commercial entities is one huge barrier to implementation of a disruptive technology option in high volume products. Deployment of a disruptive technology option on a schedule required to intersect a leading edge mobile SoC product, becomes an ongoing chicken-and-the-egg cycle: product will not adopt the technology without a demonstration, but the demonstration cannot happen without a product driver.

5.3.3   Cost Projection Conundrum

For products with tight cost constraints, such as commercial products competing in consumer market, product price projected early in the architecture phase of the Product Development Cycle, is the primary gate for potential adoption of disruptive technology option. However, the cost (AUC) and price (ASP) projections for disruptive technology options are subject to significant uncertainty and are often almost a guess. Consider the following:

- Cost analysis is, naturally, focused on the projection of cost of the product in high volume manufacturing, when most of the product is shipped. Cost estimate is therefore necessarily a projection of around 3 to 5 years into the future. As such, it is almost an educated guess, typically based on a mixture of quotes, models, judgments, and various other art forms. This is especially so for a disruptive technology—which does not have the historical data to base future projections on. Consequently, cost projections for disruptive technologies have to be on the conservative side.
- In semiconductor industry, cost structure of a Si or Packaging technology, is dominated by the CapEx. Hence cost projections can vary widely based on the expected depreciation schedule, equipment utilization, equipment sharing and uptime plans, etc., as well as on the various accounting models used to cover the CapEx. In addition, some of the disruptive technologies may require incremental investments in the factory infrastructure (e.g., enhanced defect control, new robots, etc.), and therefore the accounting approach applied to this infrastructure investment (e.g., is it amortized across all products or just the ones using the disruptive technology option) can also be a significant variable. Consequently, CapEx estimates for disruptive technologies tend to be on the conservative side.
- Cost is often driven by Yield—and projecting yield several years into the future —especially for disruptive technologies—is difficult. Si industry has over the years evolved quite a sophisticated set of methodologies to measure and model yield and yield learning. That is, the Yield and Yield Learning Rate Models have been exercised and demonstrated in Si domain over several decades. Yield modeling infrastructure for the Packaging technologies essential for 2.5D and 3D integration, and the associated OSAT portion of the supply chain, is not nearly as evolved. Consequently, yield estimates for disruptive options that leverage packaging technologies tend to be on the conservative side.
- Price versus cost is driven by the business realities, based on the value that a supplier believes a technology delivers to a customer, and on the expected competition. These factors are hard to anticipate with confidence years ahead of the manufacturing ramp. Consequently, margin estimates for disruptive technologies tend to be conservative.

Nevertheless, despite the uncertainties, anxieties and inexactitudes, cost projections must be made, as this is an essential factor driving architectural and design decisions.

5.3.4   *Risk Conundrum*

Definition of "risk" associated with use of Disruptive Technology options is the fuzziest, and very much based on gut level judgments. Consequently, any new and disruptive technology is perceived as high risk versus the evolutionary incumbent technology, by virtue of lack of experience and familiarity. From a commercial product implementation perspective, overall risk is driven by the following considerations:

- Sourcing Risks, i.e., risks due to the nature of the source of a given technology. If a disruptive technology can be sourced by a single entity only—due to intellectual property, equipment set or any other constraints, the sourcing risk is considered high. This is of critical importance especially for the mainstream mobile products, which ramp up to very high volume in a short time. For business, technical and risk mitigation reasons high volume products typically prefer to be multi-sourcable. This implies that the disruptive technology candidate must either be developed or owned by the product entity, or can be cross-licensed from a third party on terms that are acceptable and that allow multi-sourcing. Hence, highly differentiated technology options offered by a single entity are typically not desirable.
- Technology Risks, i.e., the risk of technology hitting some intrinsic challenge and hence not being available at the expected schedule, or performance, or cost point. This could be, for example, a manufacturability or reliability issue discovered at some (later) point in the technology development cycle. Historical example indicates that these incidents are "going-out-of-business" propositions and hence are often weighed very heavily during the risk assessment performed early in the Product Development Cycle. Hence, just a perceived technology risk may be sufficient to disqualify a candidate technology option.
- Cost and Yield Risks, i.e., the risk of a technology failing to come up the yield curve fast enough to meet the expected price and volume targets, resulting in supply shortages, and/or missed cost windows
- Integration Risks, i.e., risk that integration of a given disruptive technology with the rest of the ecosystem, in either design or manufacturing domain, would pose incremental challenges and issues.
- Market Risks, i.e., risk that a differentiated product, based on some disruptive technology option, is not embraced by the market, and the company is left with an orphan product.
- "Unknown Unknowns Risks", i.e., something unpredictable precipitated by the use of a disruptive technology option, and so on.

Hence, an assessment of risks associated with a disruptive technology option, conducted early in the architecture phase of the Product Development Cycle, naturally tends to err on the conservative side. And consequently, mainstream products tend to be averse to adoption of disruptive technologies—unless forced to do so by market and/or competition.

Note that High Tech business is, by definition, all about managing risks, so that semiconductor companies that deal with mainstream products are certainly capable of mitigating risk. The thing that makes mainstream products—and especially products that have a dominant market position—particularly sensitive to risk (= probability of failure), is not so much the fear of failure, but the liability associated with a failure (=cost of failure). In mobile space, a failed product could literally cost the company $\sim \$1B+$. On the other hand, emerging products naturally embrace disruptive opportunities more readily because the liability associated with a failure is smaller.

5.3.5 Benefit Conundrum

In general, if a given disruptive technology offers some advantage over an in-cumbent technology—such as form factor (smaller and/or thinner), performance (faster) or power (longer battery life)—the challenge of how to monetize this advantage.

- Monetizing the Benefit: In principle, disruptive technology advantage for a product can be used to either command a higher price, or to enhance a market position, or a bit of both. However, the advantage of increasing market share is not necessarily there for a company that already dominates a given market. Similarly, saturating markets that have reached a stable low growth rate, typically do not tolerate ASP increases. That is, there are scenarios where technology benefits do not necessarily translate to business benefits—depending on the nature of product, market, or company.
- Monetizing the Cost: On the other hand, implementation of a given disruptive technology may result in enhanced productivity and/or better intrinsic cost structure. A product using this type of technology option could then choose to either enhance its margins, or to reduce the ASP. For a dominant producer in a saturating market this would be especially attractive.
- Strategic Benefit: Implementation of a given disruptive technology may also be strategic—as in, for example, anticipating an inevitable paradigm shift at a most opportune moment, or preemptively adopting to some future constraint, etc. This could result in a strategic corporate benefit, rather than a benefit for a given IC product

Thus, a disruptive technology whose value proposition results in reduced costs is likely to be more compelling to products that are already established in their markets, whereas a value proposition that results in better performance is likely to be more attractive to an emerging product.

5.4   Observations and Opinions

Plenty has been written about the evolution of the so called More-Moore scaling versus the revolutionary opportunity of the so called More-than-Moore integration, and the intent is not to repeat any of it here. In fact, it is believed that within the foreseeable future the industry will pursue both paths, and that More-than-Moore opportunities are "as–well-as" rather than "instead-of" More-Moore scaling. Some products—e.g., Servers—may be biased towards MM scaling, whereas other products—e.g., mobile—may leverage the MtM opportunity. Time will tell. The intent of this section is to capture some subjective opinions and insights, based on author's many years of experience.

5.4.1   Product Sector Drivers

In the past, semiconductor technology roadmap was almost monolithic, and driven by sustaining Moore's Law (More-Moore scaling), dominated by digital products. It is believed that the future will be characterized with more divergence in technology trends, with forks in the roadmaps, driven by different product classes, and types. The chart in Fig. 5.1 showing market share versus the expected growth rate illustrates the point: recent R&D efforts were dominated by large digital SoC's, good for Mobile and PC space, but future development effort should feed the growth markets such as wireless and IoT. Specific considerations and opinions—with focus on 2.5D and 3D integration technologies for the next few years—include the following:

i. *Servers and Hi End PC's*: This sector is driven by *performance* and as such it is expected to continue favoring single chip SoC implementation favoring More-Moore technology scaling. However the path to higher performance is through higher levels of parallelism, i.e., CPU chips will have an increasing number of CPU clusters per die—currently 10's and growing to 100's. With this path to higher performance, and with the saturating ability to shrink dimensions, the CPU die may bloat out of a reticle size, and may therefore have to adopt a Split Die implementation, with a Si Interposer integration of some kind, in order to manage the costs. Once the mono-die paradigm is broken, there may be further opportunities for clever sharing of chip utilities (clocks, higher level caches, power and thermal management, maybe some I/O functions, photonics interface) across multiple die in a package by leveraging "active interposer" technology. Furthermore, the interface to memory hierarchy is vital for performance, and having an option for something larger than on-die memory, but closer and faster than on-PCB DRAM, may prove to be attractive. In any of these scenarios a 2.5D SiP solution with high density interconnect—and possibly including more than just wires—is likely to be an attractive option.

ii. **IoT** (industrial, smart home, etc.): on the other end of the spectrum, the IoT devices (NB: not the plumbing or the computing cloud and the associated software that this will precipitate—just the end device itself that enables the connection) are expected to be driven by *cost*. Specifically, at the highest level of abstraction IoT devices have to include (a) sensing, (b) processing, (c) connectivity, and (d) power management functionality. It is expected that with the tight cost constraints the trend will be towards specialized technology "clusters" optimized for each functionality, rather than a monolithic SoC solution implemented in a single technology. Thus, the sensing functionality will probably be dominated by MEMS technologies, and will trend towards internal integration with some local processing capability to enable a simple digital interface to the rest of the system. The processing functionality, motivated more by power than performance, will be implemented in an advanced digital technology. The connectivity capability will be implemented in a "big D— small A" type of analog technology that enables desired wireless connectivity at minimum cost and power. The Power Management functionality will require analog technology—perhaps more of a "small D—big A" type. Each functionality will be pushed towards a different technology to realize the best cost sweet spot, resulting in small die sizes—of the order of few mm on a side, or less. The challenge therefore moves towards integration of these "chiplets." It is believed that this will generate opportunities for 2.5D and 3D type of interconnect technologies. Specifically.

- Overall system integration will require low cost packaging, potentially driving FanOut WLP type of technologies towards panel level implementation
- Integration of the digital cluster with the rest of the system will require high density interconnect capability to accommodate the desired pin count on a small die, potentially driving Hybrid SLIM types of technologies.
- Integration of the MEMS with a local processor will require wide die-to-die interconnect, potentially driving 3D TSV types of technologies.

iii. **Mobile** (Phones and Wearables): mobile sector is expected to be differentiated versus the others because, in addition to the blend of all other constraints (performance and thermal a-la compute, integration of sensors and cost a-la IoT), it is characterized by significant *form factor* concerns. Thus, this sector is expected to trail the servers and PCs down the MM scaling path in order to achieve performance and power goals, but will share the MtM opportunities with IoT due to die sizes and cost structure goals. This sector will push physical dimensions; in the x–y as well as in the z-directions. Use of SiP options for heterogeneous integration of multiple die seems an obvious avenue for the mobile sector. Furthermore, the best way to address x–y constraints is to fold the electronics and leverage the z dimension—and the best way to leverage the z-dimension is 3D stacking. Ergo, this domain will have plenty of opportunities for advanced 2.5D and 3D integration technologies. In the longer term, it is expected that this domain will push breaking down the barriers between the

Table 5.2 2.5D and 3D Technology Market Drivers—a table summary of the major expected care-abouts for various principal market segments, and highlighting the expected opportunities for given aspect of 2.5D and 3D technologies

	Product sector			
	Server	IoT	Mobile wearables	Automotive
Principal Care-About	Performance	Cost	Form Factor	Robustness
Technology Roadmap Driver	More-Moore Scaling	Technology "Clusters'	Small and Thin Die	Integrated Diversity
More-than-Moore Technology Opportunities	Si Interposer (Split Die. HBM)	Small Die Interconnect	2.5D SiP (RF and Fine Line)	
	Active Interposer (shared utilities)	Cluster Integration	3D SiP (memory integration)	Multi-Physics Co-Optimize
		Panel FOWLP (low cost platform)	PCB-Pckg-Die Co-Optimize	

PCB, Package and SI Technology "silos" and will drive co-design and co-optimization across technology hierarchy.

iv. *Automotive*: there is a lot of excitement about opportunities in automotive space—coming from both the infotainment and autonomous and assisted driving requirements. This sector is expected to be driven by *reliability, robustness, and resilience* kind of considerations, without the tight constraints on performance, power or form factor. As such, it is not expected to push either the More-Moore scaling or More-than-Moore integration technologies, i.e., it will leverage developments driven by the other sectors. It may drive co-design and multi-physics design methodologies.

In summary, the proliferation of semiconductor technology, and the divergence of the markets, is expected to drive divergence in the technology roadmap. This is summarized in the Table 5.2 showing major trends for each segment, and highlighting the expected opportunities for 2.5D and 3D class of technologies.

5.4.2 Mobile Sector Drivers

The intent of this section is to summarize mobile-centric "grand challenges" and to identify potential More-than-Moore class of technology solutions; a recap of much of this book.

i. *Cost*: barring some revolutionary change (invention of some new must-have functionality or phone feature), mobile market seems to have matured into an "iso-ASP" mode, i.e., increasing performance or functionality must be accommodated within a more-or-less fixed and constant ASP. With this constraint, each new generation of products implemented in successive generations of Si technology, must be implemented in a smaller die size to cover the increasing \$\$/mm^2 processing cost. That is, even though with More-Moore scaling the cost per function (\$/gate) continues to improve, the cost of Si area (\$/mm^2) of successive technology nodes (14FF, 10FF, 7FF, 5 nm) will go up forcing die size in iso-ASP market to go down. Reduced die size necessarily drives the requirement for tighter line/space specs in packaging technology. In addition, eliminating some cost by assembling multiple die in a single SiP seems like an obvious opportunity.

ii. *Power*: reducing Vdd is still the best and most obvious opportunity for reducing SoC power, and it is expected that this will be the dominant reason for continuing down the More-Moore path. However, integration of non-digital blocks will be increasingly hard. This is then an opportunity for 2.5D and/or 3D heterogeneous integration, favoring SiP type of implementation with nonscaled analog and other functions.

iii. *Chip Performance*: higher performance and/or increased functionality is enabled through more or bigger features on a chip—more CPU cores, bigger GPU, more specialized accelerators, etc. all pushing up the die size (and cost). The pressures on die size can be expected to precipitate opportunities for split die and or heterogeneous 2.5D or 3D integration. In the long run, 3D stacking with suitably fine grain interconnect can serve to reduce the wire lengths relative to 2D planar die—resulting in better power-performance.

iv. *System Performance*: managing the integration of the processor and memory is key to the overall system performance. This pushes the need for more "on-board" memory and faster PCB-level memories, or a combination of those. The workhorse of the memory world—charge storage based memory (DRAM, Flash)—is approaching the limits of scalability and solutions that use alternative storage mechanism (magnetic spin for MRAM, material resistance for PCRAM)—will have to be developed, both for stand-alone memory die as well as for embedded solutions. This requires Si level integration of different device types—not just scaling of CMOS device size and performance—and is as such a More-than-Moore activity. In addition, with or without these new memories, placing the memory chip closer to the processor is always highly desirable—from both, performance and power perspective—opening the obvious opportunities for 2.5D and 3D heterogeneous SiP integration

v. *Form Factor*: accommodating display and battery and possibly new thermal solutions within the existing device size will continue to put pressure on the component size (in all directions). In addition, new mobile devices (e.g., wearables) are likely to put more, rather than less pressure on form factor. Eliminating packages, squeezing die closer together in a single package, and/or folding the electronics to leverage the z-dimension are all opportunities for 2.5D and 3D integration

vi. ***Thermal and Stress Mitigation***: system level integration will create increasing thermal and stress challenges. The intrinsic solution for these "new" thermal and mechanical stress challenges is to reduce the power density and increase die thickness. This is contrary to directions driven by the cost-power-performance considerations, and the trend that exacerbates Thermal and Stress issues is likely to continue. Thus, mitigation solutions will be required. New materials and new integration approaches will be a part of the solution—requiring much more of system-package-chip co-design

vii. ***Power Delivery and Signal Integrity***: In spite of the Vdd scaling, Power Delivery, PI, SI have been a growing problem exacerbated by higher integration. 2.5D and 3D integration technologies result in shorter chip-to-chip wires (alleviate the SI challenges), wider interface between die (reducing high speed signaling), smaller form factor (reduce package inductance), and opportunities for better charge storage placement (alleviate PI challenges).

In summary, the list of challenges for advanced technologies targeting mobile market is growing—no shortage of issues. Whereas, More-Moore scaling addresses some of the challenges, More-than-Moore integration offers incremental opportunities. This is summarized in Table 5.3, showing major challenges, and highlighting the expected opportunities for 2.5D and 3D technologies versus those offered by MM scaling.

Table 5.3 2.5D and 3D Technology Opportunities—a table summary of the major IC challenges and the corresponding attributes of More-More and More-than-Moore technology solutions, highlighting the attractions of 2.5D and 3D technologies that are expected to drive their adoption

Grand Challenge	Technology Attractions	
	More Moore/ SoC	More than-Moore/SiP
Si Cost	• Smaller Die	• Fine Pitch Pckg for Small Die escape • Split Die and Hetero. Integration
Power	• Reduced Vdd	• Heterogeneous Die Integration • Reduced system level wire lengths
SoC Performance	• Better transistor	• 3D stacking for reduced wire lengths • Heterogeneous die Integration
System Performance	• Better storage cell	• Different storage mechanism and cell • Better system level integration
Form Factor	• Smaller transistor	• Reduce BOM via SiP integration • 3D stacking to reduce x-y footprint
Thermal and Stress	• More Multi-Physics and Co-Design	
Power Delivery and SI	• More on-Die DeCap • Faster C2C SerDes	• SiP level DeCap and/or distributed PMIC • Wider i/f and Shorter C2C Wire Lengths

5.4.3   Concluding Remarks

In general, product management gravitates toward as-late-as-possible decisions, since this mitigates product risks, whereas technology development gravitates toward as-early-as-possible decisions since this is compatible with the development runway. This absence of a timely driving product, in addition to the absence of a well-accepted industry roadmap, creates a gap in shaping R&D Engineering for MtM technology options. However—whichever way the technology leans—the key question is "WHO will do the R&D," and especially the R&D for More-than-Moore class of technologies. Clearly, R&D and Engineering Development have to precede any product adoption, so that the direction of R&D and the choices made during development will shape the eventual product implementation. There is a gap in the industry right now - no 'guiding light' for More than Moore technology R&D.

Index

R. Radojcic, *More-than-Moore 2.5D and 3D SiP Integration*,
DOI 10.1007/978-3-319-52548-8

Printed in the United States
By Bookmasters